Pelican Books
Man and the Cosmos

Peter Ritchie Calder was born at Forfar, Angus, in 1906, and was educated at Forfar Academy. He began his career in writing and journalism as a police court reporter for the *Dundee Courier* in 1922, and subsequently worked on national newspapers, becoming Science Editor of the *News Chronicle* and joining the editorial staff of the *New Statesman*. During the Second World War he was Director of Plans of Political Warfare in the Foreign Office. Since the 1930s he has been concerned to clarify the intricacies of science for the layman and to promote its peaceful use in the world, particularly in the developing countries. To these ends he has written many books and carried out missions for the United Nations, Unesco, the World Health Organization, the Food and Agriculture Organization, the International Labour Organization and Unicef.
He has twice served on the secretariat of the United Nations Conference on the Peaceful Uses of Atomic Energy, and he holds the Kalinga prize, the highest international award for the promotion of the common understanding of science.
From 1961 to 1967 he was Montague Burton Professor of International Relations at the University of Edinburgh, and during that time was raised to the peerage as Lord Ritchie-Calder of Balmashannar. He is now a member of the council of the Open University. He is married, with three sons and two daughters.

Ritchie Calder

Man and the Cosmos

The Nature of Science Today

Penguin Books

Penguin Books Ltd, Harmondsworth,
Middlesex, England
Penguin Books Australia Ltd, Ringwood,
Victoria, Australia

First published in the U.S.A. 1968
Published in Great Britain by The Pall Mall Press 1968
Published in Pelican Books 1970

Man and the Cosmos: The Nature of Science Today is
a *Britannica Perspective* prepared to commemorate
the 200th anniversary of *Encyclopaedia Britannica*.

Made and printed in Great Britain by
C. Nicholls & Company Ltd
Set in Monotype Times

Contents

1 The Nature of Science

The everlasting 'Why?'

Science is the everlasting interrogation of Nature by Man. It certainly did not begin with Aristotle, and will not end when someone produces a 'unified theory'. Even when a theory assumes the stature of a 'law', because the consistencies seem to leave no room for doubt, it is not definitive; it is waiting for the next amendment. Newton's Laws of Gravity remained widely undisputed for two centuries until the observed behaviour of the planet Mercury did not conform exactly to them. There was a discrepancy that Einstein's Theory of Relativity helped explain and, to that extent, the laws had to be modified. Science, therefore, is *present verification* without *ultimate certainty*.

The interrogation began when Ancestral Man overcame his superstitious wondering and began to ask Why? And, in the childhood of Mankind, there was no teacher to snub that curiosity by saying 'You cannot understand that until you learn all about kinetics and dynamics.' Early man noticed an object or an event and made a mental note; thus he became an observer. He looked again to make sure; thus he became a fact-finder. He speculated why, for instance, the Sun and planets seem to move in a repetitive way against the background of the firmament and in relation to the pattern of the stars; thus he became a theorist. He searched and re-searched, testing his observations; thus he became an investigator. He began to assemble his ideas and to relate what he saw to other forms of experience; thus he became a natural philosopher. He exchanged his observations and his ideas with others, who might accept his facts (from their own observations) but dispute his interpretation of those facts. Thus there developed the dialectic – the art of reasoning about matters of opinion and the discrimination of truth from error. That was the beginning of the scientific debate that distinguishes the accretion of experimental findings from hypotheses and theories based on those findings.

Since science is non-authoritarian and upholds no dogma, it is perfectly respectable and indeed essential to the dialectic that the participants in the debate, at any given moment, should have conflicting interpretations of the available evidence – provided always that the protagonists will give way to new facts as they emerge. And emerge they will from the debate itself, because it continually raises questions to which the seekers will try to find the answers.

'God is subtle,' said Einstein, 'but he is not malicious.' Or, to put it another way, Nature is reticent about revealing its complex secrets, but is never perjured. If the answer is missing, it may be that the question has not been asked in the right way. Given the right question, properly posed, Nature will reveal the right answer in the course of time. For example, Ernest Rutherford, in a lecture at the Royal Society in 1920, offered laboratory evidence that the long-range particle he had released (1919) from the nitrogen atom by bombarding it with alpha particles was a proton – the nucleus of the hydrogen atom – and was present in the nuclei of all atoms. The observed reactions of the proton, which has a positive charge, were, he said, consistent with the conclusion that there is another particle with no electrical charge. Was there such a particle? In 1932, his colleague at the Cavendish Laboratory, James Chadwick, came through with the answer. He provided experimental evidence of the existence of the predicted particle – the neutron.

Rutherford's approach was always that of the experimentalist. Theories, he maintained, were useful tools only insofar as they explained established facts, or data that could be so established. In suggesting that the proton would be linked with another particle, he was, in effect, saying, 'I have the bolt. Find me the nut.'

On the other hand, Dirac, the experimental physicist, in 1928, postulated conditions that did not seem to make sense, the so-called negative energy states. His theory went beyond the known facts and encouraged research to find new data to support or deny his assumptions. C. D. Anderson (1932) discovered evidence for the positron, an electron with a positive charge – the sort of contradictory particle which Dirac's theory had suggested.

The outward and the inward eye

'Science' and 'scientific' have become overcharged terms because they have been given an arrogance (often by the scientists themselves) or a huckster quality (by mendacious modern advertising) or a magical quality (because the language has become an abracadabra to the layman). When we say that we live in '*the* Scientific Age', we imply that there has been some discontinuity in human inquisitiveness; what we mean is that there has been an intensification of inquiry and that there has been an acceleration of discoveries that have confirmed or disproved theories or have advanced knowledge. If our means of knowing give us greater precision and assurance, we can push more confidently and more quickly into the unknown: we can check our speculations by experiment.

This process is advancing apace today. Science has fed reliable information to technology, and technology has reciprocated by providing science with ingenious precision instruments. The new instruments have extended the range of the human senses and provided a speed of reaction and an accuracy beyond human limitations. The photo-electric cell is tireless compared to the eye; the microphone more sound-faithful than the ear; the film, the sound tape, or the video tape more retentive than the memory. There are machines a million times more sensitive than the nerve endings by which we feel, and the computer in its calculations and its assimilation and recall of 'bits' of relevant information can excel the logical (as distinct from intuitive) processes of the human brain.

Science depends on the power of observation, and observation means contact with external events or natural phenomena through the senses. When (as happened with the Schoolmen) science is regarded merely as cerebration or introspective thought-process, it becomes stagnant. It ceases to be a wellspring replenished by the rains of external experience.

When, therefore, we speak of 'modern science' and date it

from Francis Bacon (1561–1626), all that we are saying is that he rescued science from obscurantism, i.e. cerebration without imagination, and reinvoked the importance of sensory information. Bacon's time was the Age of the Eye, which men of science shared with the Renaissance artists. Indeed, both the man of science and the artist were embodied, supremely, in Leonardo da Vinci, who had both an outward and an inward vision. He could express what he saw in exquisite colour or form, or he could perceive relationships that he could transfer to mechanics, architecture, aerodynamics, or chemistry. His observations ranged widely, through anatomy, geology, geography, and botany.

When we talk of the Dark Ages, through which Leonardo flamed like a comet, we forget that in those same centuries China was having its golden age of science; the Hindus were producing experiments and practices that make the *Susruta* a scientific encyclopedia of which much is still valid by modern standards; Arabic mathematics and medicine were flowering and, half a world away, Maya–Aztec science independently had produced its own calendar. Even in the fringelands of darkness, shrouded in bigotry by the Schoolmen who corrupted the philosophies of Plato and Aristotle, there were candles in the dark; there was the first European medical college (at Salerno) supposed (perhaps allegorically) to have been founded by Elinus, the Jew, Pontus, the Greek, Adale, the Arab, and Salernus, the Latin. St Augustine (A.D. 354–430) had dominated the thinking of the centuries with his injunction 'Go not out of doors. Return into yourself; in the inner man dwells truth.' There were some, however, like William of Occam, John Duns Scotus, and Roger Bacon (all Franciscan friars), who had 'gone out of doors' and had insisted upon observation and experimental proof. This was defiance of authority in an age when men of learning could seriously argue that eggs that were most nearly round would produce cockerels. This they would justify on the grounds of Platonic 'perfection'. They insisted that all heavenly bodies must be perfect spheres, moving in perfect circles; the male sex was more perfect than the female; the hen, being the imperfect sex, could not lay perfect spheres, only pointed ovals; but some eggs were less imperfect than others – nearer the ideal round – and from these, cockerels would hatch.

To go into the henhouse to see what actually happened would have been an affront to Reason.

To establish the historical continuity of science, without which we cannot fairly enter the twentieth-century debate (because we shall not have determined, nor distinguished, the characteristics of the natural sciences we are discussing), we have to look back to the childhood of Mankind when the questioning first began for each person and each generation did not have to make discoveries afresh; they were shared and passed on as acquired experience; they were accumulated as a common fund of information; as time went on, they were set down on tablets and in books.

Astronomy was one of the earliest of the sciences. It was the recognition and codification of uniformities observed in nature. Early man saw the sun rising in different positions at different times, but always on the one horizon. He saw it set, but always on the opposite horizon, and so he recognized the eastern rising and the western setting. Given any fixed point of reference – a pillar of rock, a tree, or a pole – he noticed that the shadows that moved round it were longer in the morning and the evening, shortest when the sun was highest in the heavens at noon. He acquired a sense of timekeeping because the shortest shadow conveniently divided his working day into morning and afternoon, with the heat of the noonday sun to emphasize the division. In the wide open spaces where science began, the pageantry of the stars followed a consistent pattern in which the star clusters, the constellations to which people gave symbolic names, appeared to change their positions as the night progressed, just as the solitary sun did in the course of the day. In the northern hemisphere the observer would notice that one star (the Pole Star) would always be seen above the same point on the horizon, in the same place at sunset and at sunrise. He would notice that, as the night passed, the other stars revolved about and above the fixed star from east to west, and the division of the night (midnight) would be when a cluster rising at sunset on the eastern horizon and setting at sunrise was directly above the fixed star. So, to the 'shadow clock' of the daytime he added the 'star clock' of the night. With these physical observations he associated biological time. He (or more likely 'she') recognized the seasons and the rhythm of germination and gestation, and correlated these in the folk calendars or in practical experience.

Reasoning from the records

The regularizing of this timekeeping (by the day, the month, and the year) became a question of recording. This involved facts beyond the day-to-day experience of those preoccupied with cultivation and beyond an individual lifetime or a generation. To ensure the continuity of the records there had to be some sort of succession. The record keeping came to be vested in the priesthood. (It does not matter whether the priests became tally clerks or the tally clerks became priests; the important point is that the temples became the registries of observations across the centuries.)

Careful records extending over 360 years were discovered at Ur. These show the accumulation of information that was the basis of an advanced astronomy about 5,000 years ago – roughly when Bishop James Ussher (*Annales Veteris et Novi Testamenti*, 1650) would have had us believe the Earth was created. The records show that the Chaldean astronomers, without the help of the precision instruments now at our disposal, had worked out the length of the solar year as 365 days, 6 hours, 15 minutes, and 41 seconds – only 26 minutes, 55 seconds longer than the modern value. That functional knowledge was socially valuable. They also discovered the cycle they called the *saros*: recurrence of eclipses of the Sun. By analysing observations recorded by generations of priests, the scholars of Ur noted when the Earth, Moon, and Sun were in line, and that precisely 18 years and 11⅓ days later they would be in line again to an observer in the same position. So the priests could predict to the year, the day, the hour, and the instant when the Sun would be eclipsed. This was an important scientific fact: what some would call 'pure' science because it had no obvious practical uses. It did not tell people when to arrange their meal-times or sow their crops, for example. It did, however, give the priests apparent powers of divination.

Concurrent with time measurement there developed land measurement. Flooding rivers swept away natural landmarks that divided one tiller's ground from another's; so some method had to be devised by which boundaries could be redefined every season. That was the origin of practical geometry. The Egyptians invented a method that gave a right angle by knotting a rope in lengths of three, four, and five cubits (the unit being the length from the tip of a man's middle finger to his elbow) and pegging it on the ground at the knots. From such eminently practical devices the Greeks later produced their logical systems of geometry – although Chinese prints of 1000 B.C. show that the Orientals anticipated the famous theorem which tradition associates with Pythagoras (that the square of the longest side of a right-angled triangle is the sum of the squares of the two other sides). Excavations of clay tablets of Tell el Harmel, near Baghdad, showed that much in Euclid's *Elements* was anticipated by 2,000 years.

Numbering was similarly an exercise in convenience – counting the ten fingers (decimals); notching a stick; adding pebbles; stringing the pebbles as beads; or rigging the stringed beads in a frame, as in the abacus, an ancestral computer. Long before the Greeks, the Babylonians had a sophisticated form of numbering. Their numerals for arithmetic calculations ran from 1 to 60, 61 to 3,600, and so on – sexagesimal numbers. This was consistent with their picture of the heavens in which the twelve Zodiacal signs each occupy one-twelfth of the firmament, turning from west to east, while the Sun and planets move from east to west. The Sun takes approximately 360 days to complete the circle, thus divided by the twelve signs. On this the Babylonians based their measurements, which we still accept as the degrees, minutes, and seconds of our longitudinal division of the Earth's surface and the hours, minutes, and seconds of our clocks. Thus emerged the angular arithmetic the Greeks adopted as the basis of the geometry they rated so highly. ('Let no one enter here who knows no geometry' read the sign at the entrance to Plato's Academy.)

Babylonian astronomy served as an instrument of prediction. The tablets that have been deciphered correspond uncannily with present-day nautical almanacs, which we would regard as ready reckoners rather than as books of learning. For the purpose of forecasting it was necessary only to observe accurately and record the movements of the celestial bodies and not to speculate about how they worked. Such things were god-ordained and it would be blasphemy to question them.

On the other hand, the Greeks insisted that they were philosophers, not prophets. They regarded intellectual reasoning as the highest attribute of man. Actual observations were incidental to the unchanging principles they sought to ascribe to them. Mathematical axioms were more important than navigating a ship or any other practical purpose to which they might be put. According to Plato (*c.* 427–347 B.C.) in *The Republic*: 'Every soul possesses an organ, the intellect, better worth saving than a thousand eyes because it is our only means of seeing the truth.'

Natural philosophy

Nevertheless there was in Greece one, Anaxagoras (*c.* 500–428 B.C.), in whose works we can identify what we now accept as the scientific method. Not only did he transmit his teaching orally to students who came to 'sit at his feet'; he gave laboratory demonstrations, prepared a written account of his physical theories, and published it as a treatise on natural philosophy. Socrates turned to Anaxagoras's book to learn the laws that governed the universe.

Anaxagoras asked the questions we still persistently ask 2,500 years later. What is the nature of matter? What are the fundamental properties of the substances that make up the world around us? Is matter discrete or is it continuous? Must it all be perceptible to the senses? What laws govern the changes matter undergoes? What keeps objects together and endows them with their properties?

In pressing these questions Anaxagoras reconciled what one might call the Babylonian method of strict and detailed observation with the method of logical analysis being formulated in the Greece of his time.

He meticulously observed the phenomena of the whole physical world. When he could not satisfy himself by passive observation he would actively intervene; he would experiment. For example, he recognized that gases tend to rise above solid matter – like steam from a cauldron or smoke from a fire – and he identified air as a gas. His view of the world as a flat solid seemed to him to demand that it be supported underneath by air. He knew that air could exert pressure – as in gusts of wind – and he deduced that, although air appeared insubstantial, it should resist a large impressed force. Anaxagoras demonstrated this in public by inflating a wineskin and twisting shut its neck until the compressed air made the skin firm and hard enough to sustain heavy weights. He carried out other public experiments with the *clepsydra* (water clock), a container with small holes in the bottom and an opening at the top closed by a stopper. As long as the top was stoppered the water could not escape, but once the top was uncorked the water dropped out to serve as a measure of intervals of time. Anaxagoras stoppered a clock full of air, plunged it into water, and showed that no water entered the holes at the bottom, contending (as modern physicists do) that air pressure in the container barred the water. He also concluded that in air, the immediate dripping as soon as the stopper was removed was due to air flowing round from the bottom to the top to make room for the escaping water. He missed the significance of atmospheric pressure, apparently being unable to conceive of air (a gas) pressing downward when its nature was to fly upward. But he had the right experimental approach.

To this experimental and observational method he added logic – assembling his observed facts into patterns involving generally accepted axioms and modes of deduction that could relate antecedent and consequent events. Greek thinkers were beginning to regulate this intuitive concept that was to find elegant forms in Plato's dialogues and Aristotle's treatise on formal logic. What distinguished Anaxagoras from his contemporaries and his Platonic successors as the anticipator of the scientific method was his insistence that observation and logic were equally important and indispensable to each other. He neither allowed his physics to become pure mathematical abstraction (or introspective fantasy) nor to be dictated entirely by appearances as perceived by the senses.

From his direct observations he could see the transmutation of matter and, at the same time, its indestructability. The growing tree extracts matter from the soil and is nourished by rain from the clouds. When cut down it becomes the burning log. Part of the log rises as smoke into the clouds, whence rain descends into bodies of water in which salts and earth-precipitates form. Part of the log remains as ashes which crumble into earth and nourish plants that in turn become food for animals. To Anaxagoras it seemed that in these transformations the matter of the log eventually would pass through every state of every substance in the material world, and indeed, the universe (we can say that every breath we take contains a molecule of oxygen that was exhaled by Julius Caesar).

This idea of the conservation of matter led Anaxagoras to accept infinite variety and continuous change in the structure of matter as a universal process, and he tried to explain it in stable categories and to find the theories to justify it.

As scientific history shows, discovery depends on the Man, the Method, and the Moment. The Man may have the insight and he may have the Method, as Anaxagoras had, but it may not be the Moment because the tools and techniques may not exist. (Leonardo's aerodynamics had to wait for the internal combustion engine to confirm his theories and, incidentally, to show that Anaxagoras's air could support heavier-than-air machines.) To differentiate between elements and compounds requires careful and precise observations. Thousands of years after Anaxagoras, chemists used the refined methods and equipment of the nineteenth century to establish ninety-two elements as basic to all chemical compounds; but the Periodic Table now has been extended to include the transuranium elements. Today atoms are no longer regarded as elementary particles – elementarity is therefore a point of reference at any given time.

Anaxagoras observed only limited processes of decomposition, and concluded that an infinite set of ultimate substances existed, the elementary particles being what today might be called molecules. Much later, when it was learned how to break the bonds that held individual molecules together, atoms appeared to be the elementary particles. Atoms in turn were stripped down to yield their nuclei and electrons. With higher expenditures of energy, the nuclei can be broken up into the fundamental particles of today – how 'fundamental' we shall have to examine later.

Anaxagoras chose 'life' as his criterion of elementarity. Again we are anticipating the present-day argument about what is 'life'. The distinction between the living and the non-living has always been difficult to draw, and every attempt to understand the physical universe has had to grapple with this difficulty. Anaxagoras produced a theory of the Unity of Matter that did not require two separate assumptions – living and non-living. According to him, one of the inherent and eternal properties of all matter, animate or inanimate, was the organic nature of its basic constituents, his 'uniform substances'. This notion of living elements was most convenient because it evaded the necessity for invoking forces that entered into science neither then, nor later. It enabled Anaxagoras to construct his molecular concept of the Cosmos and to adopt *nous* ('Mind' or 'Reason') as the source of order in the universe, for which Aristotle commended Anaxagoras.

Order out of chaos

Because Anaxagoras's concepts have a relevance to contemporary debates on cosmology and the structure of matter, they are worth examining. He felt that he had to account not only for general change (and at the same time the eternal conservation of matter) but specifically for the emergence of an infinite variety of objects from any given object (e.g. from a burning log). He reasoned that the only way an infinite variety of stable entities can exist in a finite object is for the entities to be infinitesimal in size, because an infinite number of entities can be contained in a finite volume if they are infinitely small. He was therefore using (2,500 years ago) the term 'infinitesimal' in precisely the sense that it has been used since the seventeenth century: magnitudes whose measure is larger than zero but smaller than any arbitrary small number.

He also was the first on record to realize that it is not enough to determine the nature of material components that make up the universe, but that it is also necessary to provide an explanation for the motion and change that matter undergoes. He did not simply say, as other classical Greek philosophers did, that matter behaves the way it does because it is its nature to behave that way.

He sought to find a unifying principle which would account for the behaviour of each and every part of the universe for all eternity. He was saying that the goal of dynamics is the discovery of one 'law' that governs all phenomena of physical reality. He was trying to find a principle that would fulfil the following conditions: (1) it must be one not many; (2) it must account for all motion and change of state that takes place and must suffice to explain and give direction to all physical change; (3) it must be thoroughly logical and rational, and devoid of any mythological, capricious, or theological character; (4) it must account for the marvellous order that seems to exist in the universe and for the harmonious functioning of all parts of the physical world.[1] Postulating that such an entity really exists, Anaxagoras gave it the name *nous* – 'Reason' or 'Mind'.

To him this was not just poetic or anthropomorphic; it was the logical extension of his understanding of living matter. In modern terms: we observe the cellular structure and the convolutions of the human brain and infer that it is constructed of atoms, but we distinguish the anatomy and physiology from the thought processes which, somehow, those atoms generate: the mind. Just as the mind of each living being pervades his whole body, Anaxagoras's cosmic Mind pervaded all Nature. This was consistent with his theory of the composition of matter from living elements. *Nous*, or absolute Reason, in his system admits of no disorder and controls all things. The universe thus was an infinitely great structured organism – not simply a mechanism involving forces which we now can recognize, but which Anaxagoras did not conceive.

To him the matter of the universe lay muddled in chaos for aeons. Then *nous* took over. The first step separated heavier from lighter matter by churning. He had observed that eddying liquids or gases draw to their centre objects denser than the medium. Ships are sucked into a whirlpool at sea; dust is pulled into the vortex of a windstorm, houses and trees into the eye of a cyclone. *Nous*, he said, turned the handle of the cosmic churn, and out of the universal random mixture brought most of the

1. Daniel E. Gershenson and Daniel A. Greenberg, *Anaxagoras and the Birth of Scientific Method*, Blaisdell Publishing, New York, 1964.

dense matter of the universe to the centre, leaving the air and gases around. This was only a rough separation that produced a gross structure on which refinements proceeded after the churning had stopped. On Earth geological formations were created, vegetation arose, animals and humans came into being; civilized living to him was an extension of *nous*'s manipulation of molecules. The paradox was that *nous* should have been the perfection of the process, but Anaxagoras had to assume that it was the beginning. This, as we shall find, is akin to the quandary of present-day cosmogonists.

What may seem disproportionate attention to Anaxagoras in this introduction has been given as a reminder that what we are so often disposed to call 'modern scientific method' existed in its essentials 2,500 years ago. It went astray because of the Platonic enthronement of the intellect that subordinated sensory observation and promoted the abstract at the expense of the practical. (Experiment was popularly held to be for rude mechanics and not for the intellectual aristocracy.) This carried over into the age of the Churchmen and the Schoolmen, many of whom sought to discourage inquisitiveness as a threat to belief and dogma.

The scientific method

Sir Francis Bacon can be fairly said to have resurrected the scientific method by insisting on experimental investigation. He himself was scarcely a laboratory scientist, although he died as a result of an experiment (anticipating refrigeration): at the age of 65, he went out to stuff a goose with snow and contracted pneumonia. His writings (*Advancement of Learning*, *Novum organum*, and *New Atlantis*) spell out what now are widely accepted principles of modern science, rejecting the *deductive*, or thinking-off-the-top-of-the-head principle, in favour of *inductive*, or take-off-your-coat principle. He insisted that the man of science must *observe* and choose his facts; he must form a *hypothesis* that links them together and provides a plausible explanation of them; and he must carry out numerous checks or repeated *experiments* to support or deny his hypothesis. Trained as a lawyer, Bacon applied to science the laws of evidence and the burden of proof.

The scientific method, as ideally practised today, is an exacting discipline, demanding that the man of science lay aside all his preconceptions. As Claude Bernard enjoined his students, 'When you enter the laboratory, put off your imagination, as you take off your overcoat; but put it on again as you do your overcoat when you leave the laboratory. Before the experiment and between whiles, let your imagination wrap you round; put it right away from you during the experiment itself lest it hinder your observatory power.'

The man of science strives to attain the perfection of reducing everything to precise measurements. He will doubt the very sensory data that constitute his observations, often relying on sensing devices that appear more reliable and sensitive than he. Sharing the beauties of the rainbow with the poet, he also will spell it out in angstrom units (of wavelength). His olfactory nerves and tastebuds help him enjoy the bouquet and flavour of a wine, but his stern mistress, Science, expects him to express it by chromatography as aromatics. He may revel in beautiful music, but to make sound more scientifically meaningful he will reduce it to pitch, frequency, and amplitude of sound waves. He observes that the upper limit of human hearing is about 20,000 cycles per second while his instruments seem to detect frequencies of 100,000,000 cycles per second. His experiences of hot and cold often are inadequate to deal with the range and precision of temperature with which he may work in his experiments.

Precise measurements are essential because every experiment has to be repeatable, not only by him but by other men of science, anywhere and everywhere. He has to justify, not only by argument but by proof (empirically repeatable evidence), any hypothesis he tries to establish. Exact science is quantitative and not qualitative.

The observant reader will have noticed the laboured use of 'man of science'. This has been deliberate. The word 'scientist' did not appear in the English language nor, as far as one knows, in any language until 1840. As late as 1895 the London *Daily News* was still complaining about this term as 'an American innovation'. And H. G. Wells always preferred 'man of science' to 'scientist'. The distinction is important because the '-ist' meant the arrival of the career investigator and the development of specializations (each with its inbred terminology). Previously a 'man of science' had meant a person of learning, especially inquisitive about Nature – but not bulkheaded within narrow interests – and capable of sharing his inquiries and findings with other educated men. Even his terminology was common ground. It was descriptively derived from Greek or Latin roots: e.g. *protozoon* (1834) meant just what it said – first form of living things. Today the terms become more and more cryptic. From the point of departure symbolized by 'scientist', experimental science deviated from natural philosophy and other forms of learning. The tendency was confirmed when in 1901 the influential Royal Society of London dissociated itself from the concept of the Academy as the forum of all learning. Promoting the establishment of the British Academy to take care of 'moral and political sciences, including history, philosophy, law, politics and economics, archaeology and philology', the Royal Society assumed custody of the experimental sciences.

Another process had been going on – what A. N. Whitehead called 'the greatest invention of the nineteenth- century . . . the invention of the method of invention'. This was technology that increasingly brought science into the industrial arts and crafts. So there came about a hierarchy of science: 'pure' or academic science, seeking knowledge for its own sake; oriented fundamental science, research within a frame of reference; 'applied' or programmed science, which is research with a practical or manipulative purpose; and 'technology', which is the transfer of scientific knowledge into technical practice. Within such a hierarchy scientific practitioners might be distinguished as the Makers Possible, the Makers to Happen, and the Makers to Pay. But the growing tendency is to minimize such distinctions; indeed, there is an interrelationship so pervasive that it amounts to what cyberneticists call feedback: stimulation, response, and adjustment.

The theoretician adjusts his parameters to experimental evidence; the experimentalist responds to the theory; the technologist reacts to both in creating his 'hardware', and the 'hardware' itself stimulates the experimentalist and the theoretician. (Just as academic science would be crippled significantly without the computer, so the computer was derived from mathematical principles that the electronics engineers activated.) With its rapid accumulation of knowledge, with its aggregation of new discoveries, and with its impetuous conversion of desk-and-laboratory findings into practical innovations, the fantastic acceleration of scientific progress has arisen through feedback.

The result has been that the volume of knowledge – six million published scientific communications, increasing at the rate of half- a- million a year – has become a Niagara of information; that the number of scientists is doubling every ten years; and that science is becoming more and more fragmented into specializations, barely able to communicate with each other because of the unique language each has invented for its convenience. And 'natural philosophy' has been swamped by experimental results. No wonder, then, that the ordinary intelligent layman finds himself overawed and feels that scientists have become a priesthood, creating and conserving their own mysteries.

The object of this discussion is not to qualify the reader to join the priesthood. It is hoped that it will offer an insight into the processes of science that will help the reader take part in an open-ended debate, rather than eavesdrop on a closed seminar, and help him form his own judgements as to what may happen and where science is going.

As a reminder that the scientific method, on which all the arguments depend, is not at all intimidating, let us recapitulate with a simple example of Baconian principles:

You are interested in animals and you notice that some have long tails; some have short; and some (to wit: human beings) have no tails. (You never have seen the occasional infant who is born with one.) Some have toes; some have claws; and some have hooves. Some eat meat; some eat herbage; and some eat both. And so on.... But you notice that a whole group of them, including human beings, have one thing in common: they all suckle their young. So you classify them as 'milk-giving animals' or, if you know Latin, you call them 'mammals' (*mamma* = breast). Something else strikes you as significant: you have observed no mammals that lay eggs. From your data you say: 'Animals that suckle their young *apparently* do not lay eggs.'

That is your hypothesis, a provisional statement on which you proceed to do more research (which means search and search again). In the woods, farms, and zoos, and wherever you search, you continue to verify your impression that mammals do not lay eggs. You even find swimming animals, like whales and seals, that, unlike other sea creatures, do not lay eggs, but suckle their young. Now you are getting somewhere. Your *hypothesis* can be promoted into a *theory*, which is a well-established hypothesis that can be related to other accepted principles. You drop the word 'apparently'. But you are still not satisfied. You travel through North America, South America, Europe, Africa, and Asia, observing and comparing notes with others. Always it is the same: 'Animals that suckle their young do not lay eggs.' Now you may think you can state it as a universal '*law* of Nature'. You say categorically, 'No animals that suckle their young lay eggs'. But no scientific 'law' ever is proved ultimately, one reason being that it is not feasible to observe *all* the individual cases to which it is held to apply. Any 'law' in science is tentative, probabilistic, and uncertain; it is always subject to revision in the light of new evidence. Thus no scientific inquiry is ever complete, and you go off to the Australian continent to continue testing your theory. There you meet the duck-billed platypus, or duck mole, and find that this curious animal not only lays eggs but also suckles its young. All that trouble just to find yourself wrong in the end! This may be disappointing, but it is not disastrous. You can restate your *law* more precisely: 'No animals (with the exception of, etc.) that suckle their young lay eggs.'

A physiologist then may remind you that the human egg (ovum) is 'laid' in a nest formed by a fold in the wall of the womb. You accept this observation of internal incubation and restate your law adding the word 'externally'. A scientific law, like judicial law, can define the conditions in which it will apply. This mutability of scientific law is important: it is 'case law', not dogma. That is what contrasts the modern and medieval conception of science.

2 The Nature of Scientific Revolutions

The history of science is not a spoil heap of accidental discoveries and random ideas. Rather it is like an archaeological dig in which a potsherd is meaningful in terms of the chronological layer in which it is found; i.e. of the culture in which it has its origins. An archaeologist does not discard an artifact as a 'myth' because it carries mythological symbols; he examines it for its substance and appreciates the skills of the superstitious makers. In the same way, we cannot dismiss as 'unscientific' ideas which carry the imprint of beliefs that happened to be prevalent when the ideas were conceived.

That Johannes Kepler trafficked in horoscopes does not invalidate his observation that the planets revolve around the Sun in elliptical paths – a scientific principle so profound and verifiable that it made a complete break with traditions of more than 2,000 years. We can dramatize the impeachment of Galileo by the Inquisition and, by hindsight, reproach him for recanting his 'heresy', but the rearguard action of the clerics or Galileo's concessions to them is no more relevant to the fabric of science than is the picturesque legend that he dropped weights from the Leaning Tower of Pisa. It is much less relevant than the refusal of the professor of philosophy at Padua University to look through Galileo's telescope at the mountains of the Moon because it would have denied the perfection of the Platonic spheres. Sir Isaac Newton was a devout churchman, but he did not consider that he was disproving divine powers by defining the regulatory mechanisms. Michael Faraday was a member of a fundamentalist sect and so strictly believed in the Book of Genesis that he refused to dine with Queen Victoria on a Sunday, but, paraphrasing Claude Bernard, he took off his religion, as he took off his coat, when he entered his laboratory. Rudolf Virchow, the founder of 'cellular pathology', who rejected the doctrine of 'humours' of Aristotle, Empedocles, and Galen, was a member of the Prussian House of Deputies and a social (and sanitary) reformer – so much so that he expressed his human pathology in civic metaphors. His argument went like this: What makes a state prosperous? It is the prosperity of each individual citizen. What, then, makes the living body healthy? It is the health of each of its units, its cells. Disease, he contended, was anarchy within the living body. As he put it in his *Cellular Pathology* (1858): 'The body is a cell-state in which every cell is a citizen. Disease is merely the conflict of the citizens of the state brought about by the action of external forces.'

Or take scientific 'myths': phlogiston – the 'fiery principle' – we now know to be unsupported. But was it 'unscientific' in its time? As specified by Stahl (1697) it was extremely plausible. It was consistent with the observations of the time and it worked; it made possible substantial advances in chemistry in the eighteenth century. It identified the elemental fire of the Greeks as material. According to Stahl, metals were composed of calx, which was different for each metal, and of phlogiston, which was common to all metals and all combustibles. When calx (a metal from which phlogiston had been expelled by heating) was reheated with charcoal, out of contact with air, the phlogiston from the charcoal was held to recombine with calx and the metal was regenerated. Combustion, therefore, was understood to be caused by the phlogiston leaving the substances.

Joseph Priestley, Joseph Black (who defined the physical properties of latent heat), and Henry Cavendish were all believers in phlogiston. Priestley, in 1774, discovered what we know as 'oxygen'. He used a burning glass to focus the Sun's rays (thereby excluding any intrinsic chemical properties of heat) to heat the red calx of mercury, and found that in the 'new air' thus produced a candle burned much more brightly. He called the gas, thus identifiably different from atmospheric air, 'dephlogisticated air'. In another experiment he exploded by an electric spark 'inflammable air' discovered by Cavendish in 1766 (hydrogen). A dew was formed on the glass container. He found that the water thus formed was equal to the weight of two gases. He rejected the suggestion of James Watt (of steam engine fame) that water was not an element but a compound of the two gases. He left Henry Cavendish to repeat the experiment and explain the composition of water in phlogiston terms, i.e. that the water pre-existed in the gases. It needed Lavoisier, uninhibited by either the beliefs or terminology of the phlogistonists, to specify oxygen and hydrogen (both of which he named) as elements and water as a compound. Priestley, a Dissenting minister not given to doctrinal beliefs and a superb, objective, experimentalist, was strangely intractable. After the Birmingham (England) mob had burned his house for his un-British activities (sympathy with the American and French Revolutions) and he had settled in Northumberland County, Pennsylvania, he wrote (1800) *The Doctrine of Phlogiston Established and that of the Composition of Water Refuted.*

To say that phlogiston was a false doctrine or that Priestley was a bad scientist because he sustained it is like saying that the nineteenth-century view of the atom ('the indivisible particle') was false because we now know that the atom consists of electrons and the nucleus and that the nucleus itself consists of fundamental particles ('fundamental'? or 'particles'?); or like saying that Alexander Fleming was less than the scrupulous laboratory scientist he was because in 1928 he missed the real significance of the penicillin which he discovered: that it could kill germs *within* the living body. Every generation of scientists criticizes as 'error' or 'superstition' the genuinely held convictions of its predecessors which, by hindsight, we can recognize as tentative assumptions based on inadequate observations (or means to observe).

Proof without certainty

'Truth emerges more readily from error than from confusion', wrote Francis Bacon in *Novum organum.* Phlogiston was a commendable sort-out and its 'errors', though temporarily misleading, put scientists on the right track.

Bacon (in *New Atlantis*) illustrated also the proper attitude of science to what many called superstition. In describing the House of Salomon he wrote:

> We have also houses of deceits of the senses, where we represent all manner of feats of juggling, false apparitions, impostures, and illusions, and their fallacies. And surely you will easily believe that we, that have so many things truly natural which induce admiration, could in a world of particulars deceive the senses if we would disguise those things, and labour to make them seem more miraculous. But we do hate all impostures and lies, insomuch as we have severely forbidden it to all our fellows, under pain of ignominy and fines, that they do not show any natural work or thing adorned or swelling, but only pure as it is, and without all affectation of strangeness.[1]

This means that no true scientist will use what he knows to mystify or mislead others; nor will he corrupt, pervert, or withhold known facts; nor will he ascribe to the supernatural anything he can measure and show to be natural. But he is human and a creature of his times and of the influences of his training or upbringing. He only knows what he can prove, but he can speculate or believe beyond that knowledge.

In terms of the perennial arguments about science and religion, the scientist who says 'I am an atheist' is being dogmatic and unscientific. *As a scientist* he would have to show that God did not exist. He can, and must, say *as a scientist*, 'I am an agnostic,' meaning, 'I do not *know*. I cannot prove that God exists.' But he, like anyone else, can hold any beliefs as beliefs. He can, beyond empirical evidence (phenomena he can observe) and belief (speculation, which may be modified by new data), have faith, which is implicit conviction. As has been suggested, 'Science is *proof without certainty*', while 'Faith is *certainty without proof*'. (Theologians may argue that relevation is proof, but even a devout churchman like Faraday cannot measure revelation any more than Sir William Crookes or Sir Oliver Lodge, materialists in their laboratories, could establish the evidence for the spiritualism in which they believed and which they tried to explain in terms that now might be ascribed to extrasensory perception or parapsychology.)

Beyond the corpus of knowledge that represents the varied facts of science, there are corporate beliefs that we call 'schools of

1. Francis Bacon, *Advancement in Learning and New Atlantis* (World's Classics 93), Oxford University Press, 1938, p. 273.

science'. These represent the attitudes of the practitioners of science to the evidence that is common property. The treatment of the facts is selective and, today, increasingly institutionalized by the definition and tacit or explicit recognition of more and more branches and sub-branches of science. They have their own societies, esoteric seminars, journals, terminology, and professional liturgy, i.e. formularies they take for granted. Occasionally, as in the international congresses, they reconvene in their patronymic groups – astronomy, biology, botany, zoology, etc. – like the members of a separated family going back to the old homestead for Christmas. Then they may find themselves tongue-tied because of their invented languages and will gravitate into discussion groups of their own sub-branch. Sometimes there is a deliberate attempt at the reintegration of a subject, like the XVI International Congress of Zoology (Washington, 1963) for which a special medal was struck – a phoenix 'representing the reunion of zoology from its separate specialities'.

Paradoxically, as we shall see in some detail later, the working boundaries (as distinct from the nominal) are becoming less and less distinguishable. For example, it can be argued that chemistry has merged into physics and that, in molecular biology, the main preoccupation is with atomic structure, i.e. physics. This regrouping has expressed itself in combinations: physical chemistry, biochemistry, biophysics, etc.

The absorption of one subject by another is not new. An example that illustrates how two sciences with apparently distinctive characteristics can become assimilated is to be found in the reduction of thermodynamics to statistical mechanics. The mother who uses her elbow to test her baby's bath water or the nurse using a mercury thermometer to take a patient's temperature may not see the connexion with the kinetic theory of gases. Certainly anyone who thinks of measurement of temperature in terms of a clinical thermometer (which would melt at a little above 360°C.) could not grasp what the plasma physicists are after when trying to control thermonuclear reactions at 150,000,000°C., ten times hotter than the interior of the Sun. A knowledge of elementary physics, however, would inform the inquiring layman that laboratory scientists think of 'temperature' in terms other than the responses of the volume expansion of mercury. They think in terms of the changes in the electrical resistance of a body or the generation of electrical currents under specified conditions. They are able to measure temperature in terms of the thermocouple, instead of the thermometer. If the inquiring layman pursues his studies into the kinetic theory of gases he discovers that 'heat' refers to the movement of the molecules, that boiling is the agitation of the molecules, that freezing is the restraint of those molecules, and that Absolute Zero (–273·1° C.) is taken to mean that the molecules would not be moving at all.

How sciences merge

The study of thermal phenomena goes back to Galileo and his circle. During the following 300 years a large number of laws were established dealing with special phases of the thermal behaviour of bodies. From those laws emerged the systematic interrelations that came to be called 'the science of thermodynamics'. This discipline used concepts, distinctions, and general laws that also were employed in mechanics; for example, it made free use of the notions of volume, weight, and pressure, and generalizations such as Hooke's law of elasticity and the laws of the lever. In addition, however, it embodied notions of its own, of temperature, heat energy, and entropy (disorderly heat energy), as well as general assumptions that it did not share with its contemporary science of mechanics. So it was regarded as a special discipline. It still is taught as such and its concepts, principles, and laws still can be understood and verified within its own frame of reference. Experimental work early in the nineteenth century on the mechanical equivalent of heat encouraged theoretical inquiry to find more intimate connexions between thermal and mechanical phenomena. Maxwell and Boltzmann were able to express the Boyle–Charles Law satisfactorily from assumptions that could be stated in fundamental notions of mechanics. Boltzmann was able to interpret the entropy principle as an expression of the statistical regularity that characterizes the aggregate mechanical behaviour of molecules. The statistical–mechanical interpretation of the second law of thermodynamics in 1866 marked the end of the independence of thermodynamics – although it still 'flies its own flag' in the classroom and although those who apply the laws of thermodynamics in daily practice may not know (or care) that it is a secondary science and that statistical mechanics has added it as another star to its banner. This federation would not have been feasible in 1700, when thermodynamics was establishing itself as a science in its own right. Similarly, nineteenth-century chemistry was not reducible to nineteenth-century physics. In the twentieth century, however, understanding of atoms, their positioning in the molecules, and the forces that bind them together means that chemistry can be expressed in physical terms.

Most sciences have been characterized by disagreement among different theoretical schools. This is not just the logic-chopping of the Schoolmen, as in the Middle Ages, but represents genuine debate about observed facts of Nature. Observation and experiment can and must restrict the range of scientific belief, but there must always be tentative interpretations of those facts – prevailing views genuinely held and often conflicting. This does not mean that the exponents of any of the contending views are unscientific. Indeed, in terms of the essential dialectic they are scientific; without the debate there would be no progress in science. One school may be a conservative minority (like Priestley and others upholding the phlogiston theory when the majority had moved on); another may be an adventurous minority, extending the evidence into more daring speculations; and, at any given time, there will be a majority consensus imposed by dominant personalities, challenged, but widely accepted. This becomes the norm and sets the pattern of teaching in the particular scientific subject, delineating the lines of research and the conventions of practice.

It is easy to see how the framework of a scientific discipline tends to get fixed and to exclude or ignore innovation. To be manageable a body of information (and in science it is vast) has to be 'programmed' for the students. This is done principally through curricular teaching and textbooks. Courses and texts expound the accepted theories and demonstrate support for them in successful applications, prescribed observations, and experiments. They also have to take a lot for granted. The student cannot, for example, be expected to rediscover Newton's laws for himself. Textbooks, therefore, define the laws of the various disciplines and, within the prescription, they will be found to work as reliably as a logarithmic table or a nautical almanac. The students are expected to learn and accept the glossary of terms which, as defined, will serve their purpose with precision. The conventions of the prevailing schools are built into the educational framework. This does not mean that 'it is all in the books', as the Scholastics would have insisted in the Middle Ages, when it was heresy to question the canons of the Church or the works of Plato and Aristotle, and when Vesalius had to go out by moonlight to rob the gibbets because of the anathema against dissecting bodies. It means, however, that the processed knowledge of the curriculum and the textbooks is the basis for examinations and the undergraduate student is liable to be unlucky if his teacher has unorthodox ideas, i.e. outside the prevailing school of thought. Postgraduate students are (or should be) encouraged to be more adventurous. But the undergraduate has to be grounded in the accepted: law, theory, application, and instrumentation.

This inevitably reflects itself in the professionalism of science. The scientist, by his training, becomes proficient in one of the many branches of science. Science is now so wide in its scope that the individual must define the limits of his interests and he must join a particular scientific community that will have its own coherent traditions and its own frame of reference. So he will tend to conform. This will not deter him always from making original discoveries. (How else could we have had the incredible explosion of scientific innovations in recent years?) But it will not make him a scientific revolutionary. He will be constrained by what Thomas S. Kuhn has called the 'paradigm'.[2]

The paradigm

Kuhn defines 'paradigms' as 'universally recognized scientific achievements that for a time provide model problems and solutions to a community of practitioners'. A paradigm, therefore, determines the framework within which 'normal science' works.

2. Kuhn, *The Structure of Scientific Revolutions*, University of Chicago Press, 1962.

It sets the pattern of puzzle-solving, like the squares of a crossword or the box in which a jigsaw puzzle comes, the important thing being that it confidently indicates that, however complicated or whatever clues may be missing, the puzzle is capable of solution. This leaves plenty of scope for the ingenuity of the solvers and for the invention of new ideas for solving.

A scientific revolution, therefore, results from breaking out of a paradigm, committing a breach of accepted scientific tradition. A discovery, however spectacular, or a new method of instrumentation, however ingenious, is not 'revolutionary' if it is within the paradigm; e.g. the release of nuclear energy in the explosion at Alamogordo, New Mexico, at 5.30 a.m. on Monday, 16 July, 1945, changed the course of human history and was Man's greatest scientific achievement since our remote ancestors first mastered fire; but it was not a 'scientific revolution'; it was implicit in Einstein's equation $E = mc^2$ (1905) that mass could be converted into energy and vice versa. Scientists may have doubted whether, technologically, it could be done; but, according to their framework of knowledge, it was possible.

How effectively the paradigm applied in this instance is shown by the memorandum which Peierls and Frisch sent to the British committee that, in the spring of 1940, was considering the feasibility of the bomb. Within the space of three foolscap pages they showed that if uranium-235 (U^{235}) were separated from uranium-238 (with which it is associated in ores in proportions of about 1:140) there would be a chain reaction that would merit the name of 'superbomb'. They suggested the critical size of the bomb (about 1 kilogram of U^{235}), predicting that it would produce an explosion a thousand times greater than that of a kilogram of T.N.T. They suggested how the U^{235} should be separated and how the bomb might be made. They anticipated that the radiation emitted would kill living things and would persist long after the blast.

All this could be derived within the paradigm of the prevailing nuclear physics. What is significant was that neither in this paper nor in any other British papers during the Second World War was the question of the genetic effects of radiation ever raised. That belonged to another discipline – to biology, not physics – in which Hermann J. Muller had shown that chromosomes could be altered by X-rays (1927). But that evidence had escaped the awareness of the physicists; since the paradigm from physics dominated bomb technology, probable effects on heredity never were brought to the attention of the policy-makers. Clement Attlee, the British Prime Minister who concurred with President Truman in the decision to drop the bomb on Hiroshima, said later:

> We knew nothing whatever at that time about the genetic effects of an atomic explosion. I knew nothing about fallout and all the rest of what emerged after Hiroshima. As far as I know, President Truman and Winston Churchill knew nothing of these things either, nor did Sir John Anderson who coordinated research on our side. Whether the scientists directly concerned knew, or guessed, I do not know. But if they did, then, so far as I am aware, they said nothing of it to those who had to make the decision.[3]

Lord Rutherford, who, conspicuously, had helped to build up the 'nuclear paradigm', maintained until his death in 1937 that the atom would always be 'a sink of energy and never a reservoir...'. He was misled by the experience with high-voltage accelerators that used more energy to split the atomic nucleus than was released by the split atoms themselves. According to Rutherford: 'It's like trying to shoot a gnat on a dark night in the Albert Hall and using ten million rounds of ammunition on the off-chance of getting it.'[4] When, however, his own student Otto Hahn and Hahn's colleague Strassmann, in 1938 demonstrated uranium fission, the nuclear physicists (within the paradigm) said 'But of course!' and, on accepted principles, could predict (and hence were led experimentally to produce) a chain reaction in which the neutrons from one nuclear division would produce the division of other nuclei.

The Alamogordo explosion may have been an achievement within the paradigm, but it meant that the safebreakers had forced the nuclear lock before the locksmiths knew how it worked. Their corpus of knowledge enabled them to produce results, but

3. Clement Attlee, with Francis Williams, *Twilight of Empire*, A. S. Barnes, New York, 1961, p. 74.

4. Ritchie Calder, *Profile of Science*, Allen & Unwin, 1951.

they had sharply limited insight into the structure of the nucleus, which we shall discuss later in considering the fundamental particles.

Kuhn's 'paradigm' means that a 'scientific revolution' replaces one set of propositions that have served practitioners satisfactorily (even in spite of anomalies) with a new set. The Ptolemaic system served its purpose for centuries; for working approximations it is still useful. For the planets, Ptolemy's predictions were as accurate as Copernicus's. But there were discrepancies that generations of astronomers tried to reconcile with their observations. As they tinkered with Ptolemy's assumptions of circular planetary orbits to make them consistent with a particular manifestation, they would introduce cumulative inconsistencies until the exceptions perverted the rule; the tradition became so cumbered by contradictions that it just could not be faithful to the observed phenomena. The Copernican revolution was a distinct scientific revolution because it came out of a crisis: technical breakdown of the Ptolemaic system. It removed the world from the centre of the universe and put the Sun at the centre of the planetary system. The meticulous observations of Tycho Brahe and the Laws of Kepler regulated the Copernican inconsistencies by establishing the elliptical orbits, and Newton provided the physical explanation of the 'how'. Newton, however, also created a new paradigm, one of the most comprehensive in scientific history; it stimulated generations to confirm and refine his observations and his concepts, and to pursue novel thinking comfortably backed by a tradition. (It enables an astronaut to step out of a spacecraft with courage bolstered by confidence that the Laws of Gravity have not been rescinded.) With an authority based entirely on the proven and the provable, Newton was enthroned not so much as the leader of a school, but as a kind of emperor of science. His laws ranged from planets to corpuscular matter. To many, everything seemed intelligible in terms of matter and motion. Mechanical motions explained the cause of all natural effects.

Doubts about cause and effect

The confidence thus given to cause-and-effect thinking could encourage Laplace (1749–1827) to say in *Théorie analytique des probabilités* (1820):

> An intelligence knowing all the forces acting in nature at a given instant, as well as the momentary position of all things in the universe, would be able to comprehend in one single formula the motions of the largest bodies as well as the lightest atoms in the world provided that its intellect were sufficiently powerful to subject all data to analysis; to it nothing would be uncertain; the future as well as the past would be present to its eyes. The perfection that the human mind has been able to give to astronomy affords a feeble outline of such an intelligence. Discoveries in mechanics and geometry, coupled with those in universal gravitation, have brought the mind within reach of comprehending in the same comprehensive formula the past and future state of the system of the world.

This was what Sir Frederick Gowland Hopkins, the pioneer of vitamins, called 'the lusty self-confidence' that pervaded the scientific attitude (secure within the Newtonian paradigm) of the nineteenth century. During most of that century the scientists were mainly (to use Kuhn's phrase) 'mopping-up the paradigm' by perfecting methods and instrumentation; but in the main they were trying to justify each improvement or advance in Newtonian terms. James Clerk Maxwell, whose electromagnetic theory initiated a new departure, a new school of thought, nevertheless succeeded in transforming the fundamental mathematics of his theory to accommodate the Lagrangian equations, alternative formulations of the Newtonian axioms.

Maxwell's electromagnetic-field theory offered a coherent explanation for the experimental findings of Coulomb, Ampère, and Faraday, also providing a satisfactory mathematical tool for dealing with the manifestly distinctive features of electromagnetic phenomena; but, as he realized in trying to reconcile it with the mechanical conception, it presented difficulties and raised opposition. One of the difficulties was finding a material medium in which the waves could travel, governed by Newton's laws. The universal 'ether' of Newtonian mechanics might have served, but

Maxwell, having first postulated it, dropped it. (For example, if light or radio transmission represents wave motion propagated mechanically in 'ether', then celestial observations and terrestrial experiments should detect drift in that medium; in spite of all efforts and ingeniously devised equipment, no such drift was discovered.) Although he held dutiful regard for Newton, Maxwell could not fit 'ether' into his theory, and no one who tried could. Thus Maxwell became the reluctant instigator of a crisis in the Newtonian paradigm that formed the background of Einstein's Special Theory of Relativity.

Meanwhile, other difficulties in the Newtonian school had arisen. Adams in England and Leverrier in France, working without knowledge of each other and reasoning with pencil and paper, decided that orbital irregularities of the planet Uranus – which was not behaving as Newton's laws demanded – must be due to the presence of another planet. They worked out where it must be, and sure enough, when the great Berlin telescope was turned towards the prescribed spot, there was the previously undetected planet, now called Neptune. This was a triumphant endorsement of Newton's laws, but when a similar delinquent behaviour was observed in the case of the planet Mercury no such gravitational disturbance was sustainable, and again the explanation had to wait for Einstein. Indeed, by the end of the century the lusty self-confidence of cause and effect was diminished, and scientists, with more humility, were looking without knowing it for Planck's Quantum Theory, Einstein's Theory of Relativity, and Heisenberg's Principle of Indeterminacy (Uncertainty).

Determinacy – cause and effect – had dominated scientific thinking for 250 years and other fields as well. J. Bronowski (*The Common Sense of Science*, Heinemann, 1953) has shown how mechanistic philosophy, with its sense of predestination, gave an air of fatalism to the literature of the times. It affected politics in Karl Marx's materialistic conception of history, in which the cause of social and economic change is 'powers of production', and the effect is seen as an irreversible process by which one class is inevitably displaced by another. Let it be said here, however, that while Chance is back, and Probability is accepted as a perfectly valid scientific principle (for reasons we shall discuss later), the

Uncertainty Principle does not discredit cause and effect, which is still demonstrably useful in macroscopic physics.

Another form of scientific revolution was promoted by Lavoisier when he broke the framework of the phlogiston theory and liberated chemists from the limitations it imposed on observation and interpretation.

A case of trespass

Sometimes a revolution is wrought by infiltration of a discipline from another paradigm. John Dalton (1766–1844) was not originally a chemist. As a meteorologist he was investigating the absorption of water by the atmosphere. But his researches led to the Chemical Atomic Theory. By the nature of his training and approach, he considered the mixing of gases or the compounding of elements (as in water) as a physical process. He thought he could explain the observed homogeneity of solutions by determining the relative sizes and weights of the various atomic particles in his experimental mixtures. The chemists were handicapped in their thinking because of the distinction they drew between what was chemical and what was mechanical. If mixing produced heat energy, light, effervescence or something else of the sort, a chemical reaction was seen to have taken place; but if the particles could be distinguished visually or could be mechanically separated, they composed a simple mechanical mixture. Many intermediate cases were accepted by the chemists as 'chemical'. Salt dissolving in water, or the mixing of oxygen and nitrogen in air came into this category. Dalton assumed that in the restricted range of reactions he took to be chemical, the atoms only could combine one-to-one or in some other simple whole-number ratio. This led him to determine the sizes and weights of such particles. Any reaction in which the ingredients did not enter in fixed proportions was not purely chemical by this definition. For the practising chemists, suspicious at first, Dalton's 'new system of chemical philosophy' simplified their thinking and their notation – as simple as expressing water as H_2O: two atoms of hydrogen to one of oxygen. The nineteenth century was off to a flying start, with Mendeleev coming on to produce the Periodic

Table which so clearly classified the atoms and their kinship. This was a slow and relatively peaceful revolution, although no paradigm has been supplanted without some vociferous opposition from the Old Guard – in this case Claude Louis Berthollet (1748–1822), the French chemist. As Max Planck was to say later of the reception of his Quantum Theory, 'A new scientific truth does not triumph by convincing its opponents and making them see the light, but rather because its opponents eventually die, and a new generation grows up that is familiar with it.'[5]

Occasionally, but rarely, an isolated discovery may produce a scientific revolution by outraging contemporary conventions, and of itself initiate a new chain of events – acting as a fresh spring instead of drawing on the established reservoir. Such was Röntgen's discovery of X-rays.[6]

Discovery creates a paradigm

Wilhelm Röntgen's discovery in 1895 is often described as an 'accident', but it was not simply fortuitous; it was the impact of an unexpected observation on a prepared investigator. Yet Röntgen's preparation might have excluded the observation, because in the paradigm of physics of his time there was no consensus to sanction what he was observing. Maxwell's electromagnetic theory, within which X-rays would have been permissible, was still being hotly debated, and the particulate theory of cathode rays was highly tentative. Indeed, that was what Röntgen was investigating. Sir William Crookes had proposed what he called 'the Fourth State of Matter'. He had made a tube shaped like an airship, in the narrow end of which he had fixed his cathode (negative pole). Towards the bulbous end he had fixed his anode (positive pole) in the shape of a metal cross. Between those, through a partial vacuum, he passed a current. 'Something' was beamed that made the glass glow at the end opposite the cathode. The 'something' seemed to proceed in straight lines since the cross threw a clean-cut shadow on the glass. He thought the 'something' must be light until he brought a magnet near the

5. *Scientific Autobiography and Other Papers*, Philosophical Library, New York, 1949, pp. 33–4.

6. Kuhn, op. cit.

tube and the beam was bent from its path, suggesting a stream of material particles – too small to be sensed as solid, liquid, or gas; they were a fourth kind of particles, now called cathode rays. (These produce the picture in the television tube, a modern refinement of the Crookes tube.)

Röntgen, an obscure professor of physics at the University of Würzburg, was repeating Crookes's experiment and trying to explain the behaviour of the posited particles. He had been studying their effects on fluorescent salts, which glow in response to certain light waves. As a check on one of his experiments he had enclosed the Crookes tube in blackened cardboard, shutting in all light. On 8 November 1895, he noticed that a sample of these salts twelve feet away from the tube were glowing in the darkness. There was nothing in the paradigm to account for this; it was beyond his accustomed terms of reference. When years later he was asked by the famous physician Sir James Mackenzie what he thought when he noticed the phenomenon, he replied, 'Thought! I did not think. I investigated.' That is an antiparadigmatic remark because, if he had tried to account for the glow, he would have found no authority. Instead he applied to the invisible beams (assuming that they had unheard-of powers of penetration) the methods used with light. He used photographic plates, with results that are now familiar. When Lord Kelvin, then the greatest world authority in Röntgen's own field, first heard of X-rays, he denounced them as an elaborate hoax.

X-rays produced a breakout from one paradigm into another, which was to include J. J. Thomson's electrons, Becquerel's observation of emissions from uranium, Marie Curie's radium, and Rutherford's question 'Why does radium give off rays?', with the resulting nuclear research and Einstein's $E = mc^2$. By direct succession from Röntgen, in crystallography, study by X-rays permitted inferences about the atomic structure of molecules, which eventually were to indicate a helical structure for the D.N.A. molecule (the deoxyribonucleic acid of living cells) and the positioning of the atoms that may be the code by which hereditary traits are transmitted from one generation to another.

Waiting for the moment

Sometimes a scientific revolution (i.e. the creation of a new paradigm) is postponed because the supporting technology does not exist. For instance, Sir Oliver Lodge was the Man-before-the-Moment in the later paradigm of radio-astronomy. In 1900 he was convinced that the Sun was propagating radio waves, and he set up experiments to verify it; but his radio receiver (a primitive coherer) was too insensitive. In 1931 K. G. Jansky of the Bell Telephone Laboratories was studying atmospheric changes that interfere with radio reception. He noticed a persistent hiss in his earphones as he turned his aerial in certain directions. With impressive awareness, he (like Röntgen) recognized the unexpected. Repeatedly the same thing happened; if the aerial was pointed towards the Milky Way the hissing started. He inferred that radio waves were coming from the stars themselves and that the hissing from the direction of the Milky Way was the combined effect of emissions from many stars. It was like the distant sound of a ball game or the 'rhubarb-rhubarb' of a theatre crowd. When, however, he tried to locate the 'arena' and find the 'cheerleader', he was unsuccessful; his instrumentation was still not precise enough. Under compulsion of the Second World War, with the invention of radar, extrasensitive receivers were developed that could pick up ultrashortwave emissions. Sir Edward Appleton saw the possibilities of employing the new techniques to follow up Jansky's observations. Appleton's colleague J. S. Hey started searching for radio emissions from the Sun and succeeded where Oliver Lodge had failed. He turned his improved detectors onto the Milky Way and confirmed what Jansky had deduced – that the radio hiss was coming from the galaxy. They identified the signals as emanating from particular sources in the star clusters.

Planetary radio-astronomy may be said to have begun when the United States Army Signal Corps in January 1946 shot radar pulses at the Moon and got echoes back in less than three seconds, which is the time required for a radio-wave round trip of 478,000 miles. In Britain after the war, using surplus military equipment, A. C. B. Lovell (later Sir Bernard) set up an operation at Jodrell Bank, Cheshire, to study cosmic rays by radar techniques. When cosmic rays – in the present paradigm seen as high-speed particles from space – hit the Earth's atmosphere, they smash atmospheric atoms and release secondary bursts of particles. It was argued that radar could locate those bursts in the upper atmosphere just as it could detect enemy aircraft. The observed effects on radar screens were confusing until it was realized that they represented, in many cases, echoes from meteors. That meant that the radio-astronomers (as they were now calling themselves) could detect meteors in broad daylight, by getting echoes from their ionized tracks. Thus 'blind astronomy' replaced the identification of cosmic ray bursts, and meteor chasing became a sideline of a science that could re-map the heavens.

The radio-astronomers also could record the signals of cosmic events which, in terms of light-years (one light-year being the distance traversed in a year by light or radio waves at about 186,300 miles per second), happened millions of millions of years ago. For example, Chinese observers in A.D. 1054 recorded a celestial catastrophe, a star exploded. In this phenomenon, now called a 'supernova', a body similar to our own Sun suddenly 'triggers off' and expands in flaming gases to many times its original diameter. Modern astronomers optically identify this Chinese report with the Crab Nebula, still visible in the heavens as a hot, expanding gaseous envelope – a luminous ghost of the exploded star. Bolton, an Australian radio-astronomer, discovered strong, generalized radio signals from one sector of the sky. More refined data indicated to him that the origin of the signals was in the Crab Nebula. The signals are agreed to be the continuing 'radio commentary' on the event the Chinese witnessed. In 1572 Tycho Brahe recorded a supernova of which there is no visible trace; but R. Hanbury Brown at Jodrell Bank in 1952 registered signals coming from a visually blank area in the heavens. He and others checked to show that the signals were coming from where Tycho Brahe's now invisible supernova should be. Now many giant radio telescopes sweep the heavens and giant interferometers in fixed positions sort out the signals as the stars move relative to the Earth in procession overhead. With such extensions of the senses (beyond the optical reach of the astronomer), the radio-astronomers reached out towards what some may imagine to be the uttermost limits of the universe. They produced new charts of the heavens and, in profitable partnership, directed the attention of optical astronomers to celestial objects that hitherto they had not located with precision. Gradually they extended the range until they approached what many believe to be the limit of observation of the total universe – ten billion[7] light-years. This is based on the theory that the 'frontiers' of the universe as observed from Earth may be receding faster than the speed of light and therefore that the waves beyond the limit will never reach Earth. Much of the contemporary debate on the nature of the universe depends on the growing evidence of radio-astronomy. The emergence of this new paradigm, from Lodge's intuition, through Jansky's observation, shows how a hypothesis does not die when it is unsupported, only if and when it is refuted by contrary evidence.

Rediscovery of a principle

How a paradigm and the apparently rational prejudices of a school of thought can postpone a scientific revolution is illustrated by the Ehrlich Principle. Paul Ehrlich (1854–1915) was assistant to Robert Koch, who isolated the anthrax bacillus and established it as the first germ shown to produce disease in man. Ehrlich had brilliantly refined Koch's techniques of colouring histological samples, i.e. thin sections of tissue with the cells differentially stained to distinguish them under the microscope. He had concluded that different minerals and dyes had a special affinity for specific tissues that could be used to study the effects of bacterial disease. His momentous contribution was his realization that (if cells and tissues selectively accepted dyes) dyes could be used to carry special deliveries to selected parts of the body. For example, when he injected methylene blue into an animal he found that only the nerve tissue became stained, other tissues remaining free of the dye. He argued that it should be possible to inject into the living body substances that would be deposited selectively not only in body cells but in specific germs. The result

7. Throughout this book the word 'billion' is used, as in the United States, to mean a thousand millions.

he achieved was Salvarsan (Ehrlich's 606, so-called because it was the 606th arsenical compound he tried), that would selectively destroy the spirochaetes of syphilis *within the living body*, without real harm to the body itself. The physicians, medical scientists, and pharmacologists accepted Salvarsan and kindred spirochaete killers, but ignored the principle – that a tailor-made drug could safely kill germs within the living body.

Resistance to the idea had been inbred in generations of doctors. Lister was confronted by the conviction of the profession that any chemical that killed a germ *in vitro* (in laboratory glassware) must kill the living host. In defiance of this, he sprayed open wounds with carbolic acid and established antiseptic surgery. Even this appeared to confirm the professional conviction, since this antiseptic (carbolic acid) could usefully be applied to an open wound or a superficial carbuncle or boil, but would be fatal if ingested. All disinfectants were labelled 'poison'. This attitude still prevailed in 1928, despite Ehrlich's evidence to the contrary in 1909, and seems to have kept Alexander Fleming from realizing the significance of his discovery. He recognized the 'penicillin effect' – the power of the exudation of the fungus *Penicillium notatum* to inhibit the growth of germs – and used it as a laboratory technique to isolate a specific bacterium, *B. influenzae*, by eliminating other organisms. He also used it effectively in treating surface infections, but found it to be labile (i.e. it lost its effectiveness very quickly); there were other well-tried antiseptics for surface use that were less trouble. However, he did not pursue the possibility that penicillin could kill germs within the living body. 'There was', he wrote in *Penicillin*,[8] 'an idea that the common pyogenic cocci, after they had invaded the body, were beyond the reach of all chemicals.'

The fashionable paradigm prevailed. Dr Gerhard Domagk broke out of it in the 1930s when he showed that a synthetic sulfonamide dye, prontosil red, was an effective drug. This triumphantly reasserted the Ehrlich Principle by demonstrating that, without harm to the infected host, a chemotherapeutic compound could dispose of streptococci. Within the post-prontosil paradigm it was possible to discover the important

8. Butterworth, 1946.

value of penicillin. When in 1938 Ernst Chain, at H. W. Florey's suggestion, was studying Fleming's research papers, he was actually following up another discovery Fleming had made in 1922 – lysozyme. In the new climate of thinking that had developed with the use of the sulfa drugs and the belated general acceptance of the Ehrlich Principle, he came across the penicillin data, recognized its possibilities *within the living body*, and pursued them with results that much of the world knows. Incidentally, penicillin illustrates the paradigm of professionalism: the lability of the fungal exudate that had discouraged Fleming, the bacteriologist, was what excited Chain, the biochemist; he asked, 'Why the instability?' and extracted the answer.

The intellectual quantum jump

In Kuhn's useful concept of paradigm a scientific revolution does not necessarily signify itself in spectacular fashion. By hindsight we can see that Einstein's $E = mc^2$ could explode as an atom bomb, but in 1905 it had no such implications, even if the Relativity Theory of which it was part then had been understood generally. A paradigm is a kind of quantum jump from one orbit of thinking to another. Such shifts usually are made by minds relatively free of habit-formed inhibitions. That could account for the role of younger people in the breakthroughs. Kepler was 24 when he propounded the elliptical theory that demolished the antique universe on wheels and gave birth to modern cosmology.[9] Isaac Newton was 24 when he compared the force necessary to keep the Moon in her orb with the force of gravity at the surface of the Earth ('I was at the prime of my age'). Lavoisier was near 30 when he demolished the theory of phlogiston. Dalton had reached the venerable age of 40 when he produced his Chemical Atomic Theory but, as has been stressed, his background was in meteorology; he was a neophyte in chemistry. Maxwell was 24 when he made his first major contribution to the electromagnetic wave theory. Rutherford was 27, already professor at McGill University, Montreal, when with Frederick Soddy, aged 22, he effectively created the modern theory of

9. Arthur Koestler, *The Sleepwalkers*, Macmillan, New York, 1959.

radioactivity. Einstein was 26 in 1905 when he produced four papers. Each presented a major advance: the Special Theory of Relativity; the establishment of mass-energy equivalence; the theory of Brownian motion; and the foundation of the photon theory of light. Bohr was 28 when he produced his model of the atom. Heisenberg was 27 when he formulated the Uncertainty Principle.

Lord Rutherford maintained that there was among scientists a 'flint-and-tinder age' in which they 'sparked-off' innovations, and when they had passed it they should reconcile themselves to being the pundits of their subject, or the directors of research, or the mentors of inspiration. In other words, they were advised to become the wise custodians of the paradigm on the lookout for the adventurous thinkers. His counsel was 'Never say "I tried it once and it did not work."' That was the advice of one who nurtured fourteen Nobel prizewinners.

3 New Dimensions

Astronauts circling the globe sixteen times a day managed, though generations of cosmologists had failed, to convince the ordinary layman that the Earth is a small planet in the solar system. Hundreds of millions of people, by sound and vision, could vicariously share the experience of the space adventurers and, on the evidence of their own senses, could get on speaking terms with concepts that had been the concern of the few. (To the children, playing in their toy space suits, 'gee' became, not a rocking horse, but the gravitational pull of the Earth.)

The typical Earthling was led to modify his mundane ideas about dimensions, time, space, gravity, acceleration, and velocity. His world shrank in time and distance when he could watch an astronaut 'walk' across a continent from California to Florida in less time than an earthbound pedestrian would take to walk a mile. He visually appreciated the meaning of momentum in the launching of the rocket, and of acceleration in the build-up of speed to 17,500 miles an hour necessary to offset the gravitational restraints of the Earth and put a capsule into an orbit, prescribed by those same forces of gravity. Visually demonstrated, but perhaps more difficult to grasp, was the irrelevance of absolute velocity: how could a man saunter alongside a capsule travelling at 17,500 miles an hour?

The astronaut patrolling the gravitational fences might have reminded the Earthling of another thing: that Planet Earth does not consist only of the land surface on which he treads or the oceans on which he sails or the rocks into which he mines; it consists also of the atmosphere which seems so unsubstantial and yet is a substance sufficiently dense to generate enough frictional drag to present a heat hazard to the re-entering space vehicle. This should have reminded him that mankind lives in a space capsule –

the Earth and its envelope – which is travelling at nineteen miles per second in its orbit around the Sun.

The achievements of the astronauts and the space engineers may have reduced the world to scale, but with their spectacular testimony to scientific knowledge, technological skills, and human fortitude it is doubtful whether the Earthling, with humility, would accept the cosmological corollary that he, like his planet, is insignificant in the pattern of the universe. Nevertheless the twentieth-century astronaut is like a small boy paddling on the shores of an ocean; when he lands on the Moon he will have waded out only to the nearest rock; and when he has attained the farthest planets of the solar system, Neptune or Pluto (thirty and forty times as far from the Sun as the Earth is) he will still be no more than a powerful swimmer who reaches the offshore islands, within view of the lifeguards on the beach. The solar system is minuscular in the immensity of the universe; it is large (to us) only in terms of Earth-measured miles and Earth-determined time.

Assuming that space engineering and space medicine can overcome all the technical difficulties, how far could we, as biological creatures, travel in space? We first have to consider the present limitation of the human life-span. But given this time limit on travel, how far could we go? The astronauts have shown man to be unexpectedly adaptable to acceleration. We could, therefore, reasonably settle for a spaceship that always would be subject to an acceleration *g*, the same as the gravitational field that the Earth produces around us. We could at this acceleration reach remarkable velocities in the course of a few years, very close to the velocity of light: 186,300 miles per second. Professor Hermann Bondi has outlined what could be achieved:

Suppose we start off from here with acceleration *g* for a certain period, say, ten years of our lives. We then reverse the direction of our rockets and subject ourselves to the same acceleration but in opposite direction for a period of twenty years by our reckoning. The changeover may be momentarily disagreeable, but we do know that this kind of thing will not do any permanent harm to us. Having attained a certain speed relative to our starting point in the first ten years, we will need the next ten years of opposite acceleration to reduce this motion to rest relative to the starting point again, and then further ten years to bring the rocket to the same speed in the opposite direction. Switching the direction of the acceleration again, we will find that the final ten years will bring us back to rest on the Earth. Thus we will have aged forty years in this journey, about as much as we can conveniently spend during our working lives.

Seen from the Earth, however, we have been moving with terrific velocity, so much so that for the most of the time we have been travelling at almost the speed of light. In fact, as observers on the Earth see it, the farthest point reached in our travels turns out to be 24,000 light-years from the Earth. Of course, the people on the Earth have noted the passing of much more time than we in our travel at such high speed relative to the Earth. We come back to quite a different situation; to an Earth 48,004 years older than when we left. Perhaps few of us would like to undergo such an experience, but nevertheless, it gives one an idea of what we are biologically capable of. Thus we can in this way travel to places in space about 24,000 light-years away, about the distance to the nucleus of our own galaxy (the Milky Way), though not nearly as far as any other galaxy.

If we are capable of taking $2g$ for forty years then we could travel to distant galaxies over 600 million light-years away, and would return correspondingly to an Earth over 1,200 million years older.[1]

What is time?

This mathematical excursion into space should jolt us into thinking about time. It is obvious that the astronautical Rip Van Winkle, who in his own lifetime returns to an Earth that is 48,004 years older (if he travels at $1g$) or 1,200 million years older (at $2g$), is caught up in a time system that is a long way outside common experience and, to the layman, 'doesn't make sense'. Evidently it has very little to do with sundials, water clocks, sand-glasses, pendulum clocks, or spring wristwatches. These are all measurements of astronomical time based on the motion of celestial bodies subject to the force of gravity. For earthly purpose, greater precision has been given to timekeeping by quartz-crystal clocks, so accurate that it would take fifty-five years for

1. *Relativity and Common Sense: A New Approach to Einstein,* Doubleday Anchor, Garden City, New York; Heinemann Educational Books, 1964, pp. 63–5.

them to gain or lose a second. These depend on the behaviour of crystals which, when mechanically vibrated, generate an electric charge. In this way, quartz crystals can be used to express time fractions as frequencies that, in turn, can give greater accuracy to standard clocks by which we set our own watches, or can regulate the wavebands within which radio transmissions have to operate; otherwise we would have radio traffic chaos. High-precision quartz-controlled circuits are a key factor in Space Age telecommunication systems. Even those quartz-crystal measurements would not serve in 'clocking' a space-travelling Rip Van Winkle whose movements would involve relativistic (Einsteinian) considerations. For his clock, independent of the gravitational behaviour of celestial bodies, he would have to go to the atom. An atomic clock has been devised. An atom of cesium produces frequencies of about 9,192,631,800 cycles per second. These devices, from the sundial to the cesium clock, are for *measuring* time.

Time is a function of the occurrence of events. Between two nonsimultaneous events there is a lapse, an interval. Whether an alarm clock is set to predetermine the interlude of sleep or whether the date of a fossil is retrospectively fixed, the measurement of time involves a system of reference. There is the 'now' and the 'before' and the 'after'. In macro-chronology, we have historical time which reflects arbitrary date fixing by choosing an event – *anno Domini*, the birth of Christ; *anno mundi*, the date of Creation as fixed in Judaism; the *Hegira*, the flight of Mohammed from Mecca as used by Moslems, etc. By relating those events to the different cultural calendar systems, the dates may be converted from one chronology to another and, for example, to say that Mohammed fled Mecca in A.D. 622. Archaeology provides another form of dating by studying man's past through the things he made and interpreting the remains of his progressive cultures. Because it is concerned more with things than with events, this kind of dating is necessarily vague. It is based on the notion of general stages – cave-dwelling savagery, simple farming communities, incipient market-town communities, and the urban political state. In themselves these do not fix a time scale; for example, Neolithic (New Stone Age) culture, generalized for some European cultures as 'about 12,000 years ago', is reported to exist in primitive communities in the twentieth century.

Greater precision has been given to archaeological efforts by radiocarbon dating. Developed in the United States by Willard F. Libby, the physical chemist, it permits samples containing carbon to be dated to a confidently stated accuracy of about 200 years. The method can be applied to materials as much as 50,000 years old. Radioactive carbon (C^{14}) is formed when atoms of nitrogen are 'split' in the upper atmosphere under the impact of cosmic rays (from outer space). Like ordinary carbon, the radioactive 'twin' combines with oxygen to make carbon dioxide. This is absorbed by plants that fix the carbon in their tissues. Animals that eat those plants similarly fix the carbon (one radioactive part to roughly one million million parts nonradioactive carbon) in their tissues. After the plants or animals die their carboniferous remains continue radioactive, but computation indicates that after about 5,760 years half the radioactive carbon has undergone decomposition and the radioactivity of the material is only half as great as it was originally. It is inferred that after about 11,520 years only a quarter of the radioactivity is left and so the halving goes on until radioactivity is no longer detected. Accordingly, by measuring the radioactivity of a sample of carbon from wood, flesh, charcoal, skin, horn, or other plant or animal remains, the number of years that have gone by since the carbon was originally extracted from the atmosphere can be estimated.

In applying the method of radioactive dating, a sample of material containing about one ounce of carbon is burned to carbon dioxide which is then reduced to elementary carbon in the form of lampblack. The radioactivity is then measured (e.g. with Geiger counters) and compared with the radioactivity of recent carbon. The age is then calculated. The method has been checked by comparison with confidently accepted measurements. For example, each ring in the cross section of a tree marks a year's growth. The rings of the heartwood of one giant Sequoia tree suggested that 2,928 years (plus or minus 50) had passed since the wood was laid down. This compared well with the radiocarbon findings. In another application, a check was made of carbonaceous material from First Dynasty Egyptian tombs that had been reliably estimated by other means as 4,900 years old. Again the radiocarbon dating closely approximated.

By applying the radiocarbon method to samples of organic matter (man-made charcoal and other carbonaceous material) human camp sites in the Western Hemisphere have been dated to 11,400 years ago; a few that appear to date back 30,000 years have been found. Three hundred pairs of woven-rope sandals covered by a volcanic eruption in Fort Rock Cave, Oregon, were estimated to be about 9,053 years old. Linen wrappings of the Book of Isaiah from the Dead Sea Scrolls found in a cave in Palestine and thought to be from about the first or second century B.C. yielded radiocarbon ages of about 1,917 years.[2] These approximations are given plus or minus a few hundred years, since radiocarbon dating is relatively rough. Even so, it is seen to be much more accurate than estimates based on the epochal variations of descriptive, adjectival ('Paleolithic', 'Neolithic') archaeology.

Radiocarbon dating is eminently suitable for estimating age for the later period of geologic time (Recent Cenozoic); the rock history of Modern Man thus can be traced and his remains dated. In dating beyond 50,000 years ago, geologists, and others who deal with older fossil remains of plants and animals found in rock layers, use the so-called radioactive clock. This makes use of end-products of radioactive decay in the form of isotopes of lead (Pb^{208}, Pb^{207}, and Pb^{206}). Such radioactive elements as thorium and uranium decay into those forms of lead at well-defined rates. For example, one gramme of uranium (U^{238}) would be expected, in 4,500,000,000 years, to have decomposed to leave half a gramme of uranium while producing 0·0674 grammes of helium and 0·4326 grammes of isotopic lead. If a rock contained these elements in such proportions, it could be taken as evidence that the sample was 4,500,000,000 years old. (Thorium/Lead, Potassium/Argon, and Rubidium/Strontium ratios also are being used for estimating the ages of rocks.) Rocks have been found on Earth with ages, thus determined, of more than 3,000,000,000 years. Meteorites that have landed on Earth indicate ages of

2. Cf. Linus Pauling, *College Chemistry*, 3rd ed., W. H. Freeman, San Francisco, 1964, pp. 792–3.

4,500,000,000 years. (These meteorites are attributed to a planet believed once to have moved between the orbits of Jupiter and Mars. Broken up, probably by the tidal forces of Jupiter, its fragments are now held to be moving in the neighbourhood of the old orbit; a string of asteroids can be observed there. When they break loose they occasionally fall on Earth as meteorites. Their chemical constitution strongly suggests that they solidified under great pressure such as would be found in the interior of a planet.)[3] The evidence tends to support an estimate of the age of the Earth and the planets of the solar system at 4,500,000,000 years.

What is space?

Time is the 'when' of common experience, but there is also the 'where', that is, position in space. In the abstract, this is difficult to grasp. To the real-estate agent 'space' is what he pegs out as a 'lot' and sells. To those concerned with space exploration, 'space' extends outside the confines of this planet. We cannot ordinarily 'imagine' space; we tend to relate the concept to material objects as they appear in our sense experience. We are led to discuss space in terms of markers, of the relative position of bodies. In those terms, space is conceived in a physical context, constrained by the observation that material bodies occupy different positions. This is a most convenient notion when all the positions of bodies are referred to one body, e.g. the Earth. This is what continues to make Euclidian geometry so satisfactory for so many practical purposes. The point, the straight line, and the plane often are accepted as having a self-evident character.

This apparently logical self-sufficiency, however, has become most inconvenient for physicists. In ordinary mechanics every event seems determined by 'place' (position relationship) and 'when' (temporal relationship). To two-dimensional and three-dimensional space (geometry) was added the inescapable consideration of time, but space still could be accepted as a separate and perfectly serviceable dimension in classical physics. This reconciliation was possible because of what we now recognize as

3. George Gamow, *The Creation of the Universe*, Viking Press, New York, 1959.

the illusion of 'simultaneity', i.e. when we receive the news of an event (say, the motion of a planet) instantaneously by the agency of light, we see it happening here and now.

The faith in absolute simultaneity was destroyed by the laws revealed by electrodynamics. Maxwell found that disturbances in an electromagnetic field travel with a very definite velocity, the speed of light, whatever their wavelength. *Light is only a special case of all those disturbances which are wave phenomena*, all of which have a periodicity and a wavelength. Heinrich Hertz (1857–94) later discovered that ordinary electric disturbances – 'sparks' – could produce an electric field some distance away. This led to the detection and transmission of radio waves with their enormous variety of wavelengths. These vary from the long waves used in wireless telegraphy to the short waves used in television and radar. Corresponding to the wavelength there is the frequency, which is the number of oscillations per second. Frequencies are measured in hertz (cycles per second); e.g. Kilocycles (thousands of cycles) per second or kilohertz; megacycles (millions of cycles) per second or megahertz. (To find waves that are shorter than the shortest radio waves we go to atomic or molecular excitations, and for shorter still, to nuclear excitations.) The visual wavelengths, which we call 'light', are those that excite the atoms of the materials that form the retina of the eye. These atoms respond to various wavelengths around one fifty-thousandth of an inch. The longest normally give us the experience of red; the intermediate ones give us yellow, green, and blue; and the shortest ones, violet. Longer than visible waves – but shorter than radio waves – are the infra-red (heat-producing) waves. Beyond the usual violet, there is ultra-violet (which gives us suntan). Shorter still are X-rays, and even shorter are gamma rays from the nucleus of the atom.

What is remarkable is that this vast range of waves, of different frequencies, differently excited and differently received, all behave in accordance with the laws which James Clerk Maxwell propounded. Another feature is that waves can be bent by reflection; e.g. the radio waves that are mirrored by the ionosphere so that they can overcome the curvature of the Earth. Higher frequencies such as those needed for television do not act so conveniently. They are limited by horizon range or have to be purposefully redirected by communications satellites. They all travel, however, at the velocity of light, about 186,300 miles per second.

Consider the application in radar, the means by which radio is used to locate distant objects. A short pulse of radiation is bounced off the target, giving information about the direction and distance of the target. The direction is simple – just that in which the signal is sent out. Distance is measured by inference from the interval of time between the instant of transmission of the pulse and the instant of reception of the echo. Since one knows that radio waves travel with the speed of light, the interval multiplied by the speed of light and divided by two (half the round trip) gives the location of the target and the continuing signals, bouncing back, give the speed of the quarry.

What is space–time?

Hence we have a method of measuring distance in which one does not use a yardstick. No standard metre or standard yard is employed. What one does is to measure an interval of time and then multiply this by a constant quantity, the velocity of light. We can thus use time to express distance. This, of course, is familiar in astronomy; to avoid awkwardly big figures in discussing the distance to the stars in miles, we express distances in light-years, based on the distance (about 5,878,000,000,000 miles) that light travels in a year. We also can speak of 'light-microseconds', based on the distance light travels in a millionth of a second. This useful unit is about 300 metres or 330 yards. A 'light-millimicrosecond' would then be a thousandth of this unit, roughly equal to one foot.

In ordinary, mundane affairs, as in clocking a race or setting a radar speed trap, the factor of the speed of light is not likely to matter very much (one would scarcely fight a 'ticket' for speeding by disputing the measured mile in terms of five millionths of a second); but, when we come to talk about the nature of the universe, such aspects of time and distance become important. Once we have got the idea that time is distance and vice versa we recognize that the star we see is not a here-and-now phenomenon, but that its light has been travelling to reach us for billions of years, which means that it is billions of billions of miles away and that what we have been witnessing, in our present, was an event of the remote past. When space and time become thus inseparable, when one cannot think of one without the other, time ceases to be one-dimensional as in classical mechanics and space–time becomes the fourth dimension.

Common sense boggles at the idea of four dimensions which no one, not even the physicists and mathematicians, have been able to imagine, far less to see. All that is meant by four dimensions is that the time and space coordinates can specify an event: when and where it takes place. It is also rather uncomfortable, in common-sense terms, to think of the here-and-now as the past, as in the case of the remote stars, or as the future in terms of the Rip Van Winkle illustration. A watch or a speedometer can serve our ordinary purposes perfectly well without invoking the space–time continuum. The basis for measuring time is unchanged for practical purposes when the velocity is very small, but changes substantially at large velocities. Space–time is a concept to which we have to accustom ourselves when we are thinking of the nature of the universe, or when we are dealing with particles accelerated in atom machines until they acquire a velocity very near the speed of light.

Before one considers other implications of Einstein's Theory of Relativity, which demanded this idea of space–time, one has to consider his 'inertial observer'. In common language inertia means an indisposition to exertion or action – laziness if you like – but to the physicist inertia means something different. In physics it refers to the tendency of matter to remain at rest or in uniform motion in the same straight line or direction unless acted upon by some external force. An inertial observer, therefore, is a person who is not being accelerated, who is not 'stepping on the gas' nor being rotated as on a merry-go-round. For instance, once a traveller in a jet liner is airborne and travelling at a steady velocity, he is 'inertial'. He reads his book or pours himself out a drink as steadily as if he were sitting in his armchair in his own house anchored on its immobile foundation and may be quite oblivious to the fact that he is travelling at 600 miles an hour –

unless he looks out the window and sees the landmarks moving away from him at that speed.

'Inertial observers' are therefore points of reference in relation to an event: not the player running with the ball, but the referee, the linesman, and the spectator each looking at what is happening from different vantage.

Classical relativity

Obviously, if different observers are to make observations on the same set of phenomena, the accurate comparison of their findings must depend on agreed concepts of space and time. Each observer's results must be convertible into the terms used by other observers. It is like translation from one language into another that can convey an exact meaning provided the respective foreigners agree on their dictionary definitions; e.g. that '*vache*', in French, means 'cow' and not 'horse', or, more aptly, that '*13 heures*' means '1 p.m.' or 'kilometre' means five-eighths of a mile. Mathematicians call this a transformation from one system to another and express this by means of a set of 'transformation equations'. Just as a dictionary must not add to nor subtract from the intrinsic meaning of a word, the mathematician must use 'invariants'.

Classical (i.e. pre-1900) mechanics could get along satisfactorily with such transformation equations in what can be called 'Newtonian relativity'. Let us take a simple example: a railway track runs across the Great Plains – no curves and no hills to interfere with constant velocity. Saboteurs have mined the track and explode two bombs at different times, each at a different place. The explosions are observed by two different people – the stationmaster at a depot and a traveller on the train. They have identical watches and as the train flashes through the station, proceeding at constant speed, each notices that it is precisely 12 o'clock. Each observer is using instruments to measure distance. When the bombs go off, the stationmaster will put down two numbers X_1 and X_2 as giving the distance away and checking with his station clock he will put down T_1 and T_2 as the times of the separate explosions. The man on the train speeding towards the explosions will put down X'_1 and X'_2 for his distances and T'_1 and T'_2 for his times.

In classical mechanics, as we have seen, time and distance were two separate absolutes, unchanging and the same for all observers. If the stationmaster finds that the distance between the two explosions is ten miles, the man on the train must also so find it, and if the man on the train says that the time interval is thirty-five minutes the stationary observer must agree. Furthermore, in this frame of reference, time intervals and distance intervals between events must be the same for all observers no matter how they are moving relative to each other. This gives a very simple transformation equation by which a description of any event in a moving system can be derived from that in a fixed system. First, all the T values (time as noted by the stationary observer) are agreed to be equal to the T' values (time as noted by the moving observer). Second, to adjust the X' value of any event, that is, its distance from the moving observer, it is only necessary to multiply the velocity of the traveller by the time of the occurrence of the event and subtract this from the X value of the event, that is, the distance of the event from the stationmaster. The mathematician expresses this transformation in the following terms:

$$X' = X - vT$$
$$T' = T$$

where v is the speed of the man on the train. All that has been stated in the transformation is that the fixed clock and the moving clock keep identical time and that the distance of points on the railway track is less for the traveller than the distance of the same points from the stationmaster by the distance the man on the train has moved from the station. That, it would seem, makes common sense.

Science, however, imposes another demand – the Principle of Invariance which insists that transformation from one system to another must not trifle with a law of Nature, just because it is convenient to do so.

The mathematical transformation given above certainly takes no liberties with Newton's three laws of motion and the principles of the conservation of energy, and until the middle of the nineteenth century nobody felt called upon to question it. Then Clerk Maxwell, following Michael Faraday's experimental work on electromagnetism, produced his equations showing that all radiation (including light) is propagated in the form of waves. What he showed was that the speed of light is constant, independent of the motion of the source and of the motion of the observer. This means that if a lamp is flashed at the railway station and both the stationmaster and the passenger measure the speed of light as it moves in the direction of the train, they should both get the same result. (The same would apply if the beam originated from any star.) If we apply the transformation equation used in the 'sabotage' illustration, it would follow that the traveller would find (leaving out how he could cope with the measurements of 186,300 miles per second) the speed of light somewhat less than that calculated by the stationary observer. This is consistent with ordinary experience: if a car is travelling at forty miles an hour and a second car is travelling at fifty, it will draw away at ten miles an hour. This seems manifestly true of moving bodies and, as the nineteenth-century physicists maintained, anything that Maxwell said to the contrary must be wrong.

Then, in 1887, an experiment was devised (Michelson and Morley) to test whether there was a difference in the speed of light in different directions. If the speed of light was different for a moving observer it should have been possible to show it, by measuring the speed of light and comparing the result with the value given by Maxwell's theory to determine the speed of the Earth through space. In other words, it should be possible to detect absolute motion.

If we consider the speed of light in terms of a beam moving in the same direction as the Earth's motion and of another beam moving at right angles to this direction, it should have been possible to detect absolute motion; and if the absolute concepts of space and time were right, then, certainly, Maxwell must be wrong. Michelson and Morley experimentally measured the two beams and showed a result that meant that either the Earth was standing still or Maxwell must be right.

That contradiction of classical physics only confused the situation until 1905 when Albert Einstein proposed a solution. He started from two fundamental assumptions: the first, that all laws of Nature must remain unchanged when one transforms from a given system to another system that is moving with uniform velocity with respect to the first system – the Principle of Invariance. If the transformation equations are based on the correct concept of space and time and the so-called law does not conform to this principle then it must be amended so that it does. The second assumption (which Einstein accepted as a fact, in view of the Michelson–Morley experiments) was that the speed of light is constant to all observers. From these assumptions he went on to derive a new set of transformation equations that unify space and time: the Space–Time Continuum.

Einstein's relativity

Although it is like trying to measure one's waistline in angstrom units (of the wavelengths of light), we can consider classical space and time absolutes and relativity space–time by looking again at the 'sabotage' illustration. As we have seen, the distance between the two explosions should be the same for both the moving and the fixed observers and the same should hold for the time interval between the incidents. In the Theory of Relativity this is no longer the case because distances and time intervals vary with the speed of the observers. There is, however, a space–time interval between events that remains constant for all observers. No matter where they are or how fast they are moving, they are all 'within the law'.

As a consequence of this theory, simultaneity and distance are not accepted as invariants. If two occurrences are observed (call them A and B), one observer may insist that they happened simultaneously, another may maintain that A happened before B, and a third that B happened before A; each will be faithful to his own frame of reference. Similarly, an observer may measure a definite distance between events, but another may say that they coincided in space.

Let us be quite clear: all this squaring, multiplying, and subtracting is not going to settle a ball game argument among the referee, the linesman, and the spectator; nor has it relevance to everyday experience, because the adjustment of the transformation equation becomes imperative only when the speeds involved get close to the velocity of light. Such speeds are now obtained in the giant accelerators that impel particles of the atomic nucleus. At those speeds time and space get hopelessly mixed up and the only measure that still remains the same for all observers is a certain space–time interval.

Another equation emerged in the development of Einstein's Special Theory of Relativity: $E = mc^2$, which we now recognize as the trademark of the atomic bomb (of which there was no premonition at the time). This equation expresses the equivalence of energy and mass. It says that energy equals mass multiplied by the square of the velocity of light. It says that a moving body, whether it is a particle or the sum of many particles, gains mass (amount of matter) proportionate to its increase in speed. Conversely, as has been shown with cataclysmic violence in the explosion of nuclear bombs, energy is released proportionate to the loss of mass. In the case of the atomic or fission bomb the sum of the mass of the elements produced by splitting is less than the mass of the original atom. The complete fission of a pound of uranium would produce about 2½ million times more energy than is produced by the combustion of a pound of coal. In explosive power, the release of nuclear energy is expressed in kilotons or megatons, the equivalent of 1,000 and 1,000,000 tons of T.N.T., a chemical explosive. The energy of fission of 110 tons of uranium is one megaton. In the case of the thermonuclear bomb, the fusion of lighter atoms to form heavier atoms – hydrogen into helium – releases even greater energy than the fission of heavy elements. The process of converting 4 hydrogen atoms into 1 helium atom, which is understood to be the principal source of energy of the Sun, involves the conversion of 0·7 per cent of the mass into energy. For bomb purposes the reaction of a deuteron ('double-hydrogen') and a triton ('triple-hydrogen') to form a helium nucleus plus a neutron is accompanied by the conversion into energy of 0·4 per cent of the combined mass of the deuteron plus triton.

At the beginning of the twentieth century, therefore, there was an upheaval: the Newtonian paradigm had been found wanting.

Newton's three laws of motion could not completely meet the observed facts. For example, the law which says that the force acting on a body is obtained by multiplying the mass of the body by its acceleration can have no ultimate meaning if the mass of a body changes as it moves; there is no way of knowing which value of mass to use in expressing Newton's law. His law of gravity involves the masses of bodies and the distances between them, and in Einstein's theory both these quantities vary from observer to observer. Which mass? And what distance? Einstein, after producing his Special Theory of Relativity in 1905, set out to formulate his General Theory, to reconcile relativity and gravitation.

Since the gravitational problem is concerned with the mass of the body, he directed himself to this and to the fact that mass appears in two different identities in classical mechanics. Classically, mass is the quantity obtained when the force exerted on a body is divided by the acceleration which that force imparts to the body. This quantity is referred to as the 'inertial mass' of a body. But 'mass' also classically refers to the quantity called the 'gravitational mass' of a body in the formula for the force of gravity between two bodies. Experimentally the inertial mass and the gravitational mass are numerically equal, although the two quantities are physically unrelated. Einstein was convinced that this was no terminological coincidence; that there was a 'principle of equivalence'. Let us imagine Isaac Newton in the seventeenth century sitting under an apple tree which is growing out of the solid English earth. He watches an apple fall to the ground and perceives a profound truth: a force from the Earth is pulling down the apple, and all other bodies, with the same acceleration regardless of their mass. He calls this 'the force of gravity'. Let us now imagine Albert Einstein, in the twentieth century, in an elevator suspended in space. It is a curious elevator because there is an apple tree growing in it and its upward acceleration is 32·2 feet per second. An apple falls to the floor, which is not the massive earth attracting the apple, but a platform rushing upwards to meet it. Newton's gravitational pull *downwards* has become Einstein's acceleration rate *upwards*, and there is no way by which either observer can experimentally prove whether he is at rest in a gravitational field or moving in an accelerated system.

A straight line, we were once told, is the shortest distance between two points; but if we draw a line on a sphere – a model of the Earth – linking two points, the shortest distance along the surface will be the arc of a great circle. (This 'great circle' is now familiar to us as the shortest route from our airport of departure to our airport of arrival.)

Gauss, the German mathematician (1777–1855), had provided mathematical tests for determining curvature or flatness. If the curvature is zero, the surface is flat and the shortest distance is the shortest distance of classical geometry, but if it is positive (greater than zero) or negative (less than zero) the surface is either elliptical or hyperbolic, and the shortest distance is a different kind of 'straight' line. This work of Gauss had been further developed by Riemann, the German geometer who applied it to curved spaces of any number of dimensions.

This non-Euclidian geometry was available to Einstein when he needed to apply it to the concept of Four-Dimensional Space–Time. He assumed that the space–time construct of the universe is not flat but curved. He mathematically formulated the theory that a freely moving body (i.e. one that is not subjected to the push or pull of any other physical object) moves in a Euclidian straight line unless a gravitational field is present. If it is, the body moves in a curved orbit which is the four-dimensional, or space–time, shortest distance between any two points. By this method the gravitation field acting on a body is no longer considered as a force but as a curvature in space–time. He worked out the invariants that would apply under all transformations from one system to another. Astronomical observations of the planets and of the Sun at the time of its eclipse have endorsed Einstein's 'curvature of space'.

The black-body revolt

Another paradigm was emerging at the beginning of the century. It began with what may be called 'the black-body revolt' against Clerk Maxwell's electromagnetic-wave theory. That theory, as we have seen, itself had been a scientific revolution. According to the wave theory, light is a periodic electromagnetic vibration propagated through (etherless) space. In the case of visible radiation, the wavelength is expressed in angstrom units which are 100,000,000 times smaller than a centimetre. The red colours which our eyes see have wavelengths of about 7,000 angstroms and deep violet colours wavelengths of about 3,500 angstroms; but as Clerk Maxwell showed, radiation is not restricted to the visible range. In the radiation spectrum we have radio waves a mile long (from crest to crest) and, at the other end, gamma rays from the nucleus of the atom with wavelengths equal to a fraction of an angstrom. Then there is the frequency of radiation. (There is nothing complicated in this idea: watch a buoy bobbing up and down on the sea waves; the number of bobs in a given time gives the frequency of the waves.) In the case of visible light, the frequency (or vibrations per second) is very large: when we see violet, the retina of our eye receives about 1,000,000,000,000,000 vibrations per second.

It was this question of frequencies that showed the weakness of the wave theory as expressed by Maxwell; it did not fit all the observed data. If a piece of metal is heated it turns dull red, then bright red, and then white. As it gets even hotter it emits violet rays and ultraviolet rays. There is, however, a special case known as 'black-body radiation'. This we can understand if we consider what happens if radiation falls on an opaque surface. Part of the radiation is absorbed and part is reflected. If the surface were *perfectly* white it would all be reflected (that is why, approximating the theoretically perfect white, we wear light-coloured summer clothes, construct white buildings, or use aluminum coatings for heat-repellent suits or constructions). On the other hand, if all radiation is to be absorbed, a perfectly black surface is necessary. A black body, however, not only absorbs heat energy; it emits it: not by reflection but by accumulating it so that it gets hotter and hotter and radiates it over the entire continuous spectrum.

To reproduce, experimentally, the black-body effect, it is possible to enclose an intense heat source in a container (furnace) and make a tiny hole in it. The experimenter can measure (by a spectroscope) the components of the radiation being emitted through the hole. As the temperature of the source increases so does the amount of energy radiated and the intensity of violet frequencies. Maxwell's wave theory could deal with the first finding. It takes care of the observation that if the absolute temperature is doubled, the total amount of energy emitted per second from the pinhole will be sixteen times greater; and if the temperature is tripled, the rate of the emission will increase by a factor of eighty-one, and so on. It also agrees with the experimental finding that the wavelength of the maximum intensity varies inversely with the absolute temperature (the hotter, the shorter).

Beyond that, the wave theory could not cope with observed facts. Pushed to the limit, the theory would have produced *ultra-violet catastrophe*. (Scientists, who usually deplore sensationalism, can nevertheless indulge in sensational terms. All this portentous description means is that a black body should emit all its energy in the ultrashort wavelength region in one violent outburst.)

Every attempt to reconcile the experimental observations with the theory failed, and in 1900 Max Planck showed mathematically that as long as it was assumed that a black body emits radiation continuously in form of waves, the ultra-violet catastrophe was inescapable. He suggested that radiation is not emitted continuously, but in little discrete packets (quanta) – each 'photon' or quantum of light having its own wavelength and frequency. This was not a resurrection of the corpuscular theory of light that the nineteenth-century Newtonians had tried, and failed, to reconcile with the wave theory. Planck's 'photon' could have no mass when at rest; it could have existence only when it was moving with the speed of light; it would vanish, or become part of an ordinary particle, when it came to rest. This posited substance which was conceived to be as unsubstantial as a shadow did not satisfy Einstein. He insisted that not only was radiation emitted discretely in the form of photons, but that it continued as photons at all times. He drew his support from, and at the same time explained, the *photo-electric effect*.

Einstein's experimental support came from Heinrich Hertz's work on radio. Hertz used an induction coil to produce sparks between two metal knobs and found that he was able to get better sparks if he irradiated the knobs with ultra-violet light. The air between the knobs became electrically charged when the metallic knobs emitted electrons under the influence of ultra-violet light.

According to the intensity of the rays, or the sensitivity of the materials used, the emission of electrons can be produced by visible light as well. This is the explanation of the photo-electric cell, which changes light into electric currents – the flow of electrons. Consider what this has meant in the entertainment industry (among thousands of other applications). The photo-electric cell permits use of the sound track of movie films. The television camera (for black and white, or colour) consists of a mosaic of photosensitive cells, which, responding to the gradations or colour frequencies of light, converts them into electric signals. These can be transmitted, to be picked up by a television receiver and converted into a beam of electrons in the television tube. This beam impinges on the flat internal surface of the tube, which is coated with chemicals. Those, in their turn, convert the cathode rays back into light frequencies that the viewers can see. Or the frequencies produced by the original photons (the parcels of light that hit the camera mosaic) can be stored as untransmitted signals on a video tape, just as voice vibrations, converted into electron patterns by a microphone, can be stored magnetically on the familiar recording tape. When the magnetized patterns are reactivated by a magnetic pickup head, the original electron process is reproduced. A photon, therefore, is 'something' which punches electrons out of the atoms of a material and frees them to move. (To paraphrase the title of a television programme, 'Have Photon. Can Travel.')

The speed of the emitted electrons depends only on the frequency of the light used. The higher the frequency (or the bluer the light) the faster the electrons move. If the intensity of the light is reduced, the speed of the electrons is not affected – only fewer electrons are punched out. Einstein was able to produce a formula which can be extended to cover the behaviour of light at all times and accounts for the measured energy of the emitted electrons.

The relationship between energy and colour can be expressed

in this way: to find the energy of a photon take the frequency of the photon's motion and multiply it by the number 6·625 divided by one and twenty-seven zeros: 1,000,000,000,000,000,000,000,-000,000. (Why? We shall take Planck's word for it.) This is written as $6{\cdot}625 \times 10^{-27}$ or, in mathematical shorthand, *h*.

Planck's Constant *h* is one of the indispensables of modern science. If it were successfully challenged (and, since nothing is scientifically sacred, attempts have been made to do so) it would upset most of the present theories of the universe and undermine the paradigm of the Quantum Theory.

To most people, the Quantum Theory is as intimidating as Einstein's Theory of Relativity, of which it was once said: 'Only three people understand it – Einstein, Bertrand Russell and God.' The Quantum Theory is now the working tool of thousands of scientists (and the modern *Pons asinorum*, 'Bridge of Asses', of hundreds of thousands of students). Since it is now so fundamental to physics, from the nature of the heavens to the nature of the atom, it is advisable to get on nodding terms with it.

Quantum paradigm

Quantum mechanics has the reputation of being much more abstract than the classical mechanics that reigned supreme before it. The reason for this is that the objects of study of quantum mechanics are much less directly related to everyday life than those of classical mechanics, or than the objects of classical mechanics appear to be.

The word to emphasize is *appear* because the 'everydayness' is deceptive. For instance, Newtonian mechanics considers first 'mass points' and then 'rigid bodies' that are supposed to be composed of 'mass points'. One could take a billiard ball as exemplifying a rigid body, or, with some abstraction, a mass point. It is usually accepted, without too much questioning, that Newton's mass point has fewer properties than a billiard ball that one could actually hold in the hand. It has, unlike the billiard ball, no colour, nor can it be thought of as being warm or cold. Moreover, it does not, like the billiard ball, occupy a space, because (as Euclidian geometry insisted) a point has no extension.

This used to worry classical philosophers who occupied themselves a great deal with the problems posed by such notions as the extension (or lack of it) of a point. Such arguments can be conveniently ignored in Newtonian mechanics; one writes down equations for the extensionless, colourless, heatless mass point and interprets what one has put down in terms of a mental picture of the billiard ball or something like it.

If one wants to be rather more realistic about a billiard ball, one treats it as a Newtonian rigid body – with a lot of reservations. For one thing, the ball is not absolutely rigid. Even if it is solid ivory (or if it were made of steel, like a ball bearing) it can be compressed and distorted, given enormous forces. If we look at it under the microscope we see that it is not bounded by a smooth geometric surface. Atomic theory tells us that it consists of particles ('mass points'? or 'rigid bodies'?) with spaces in between each and all in quite violent motion. Nevertheless, what Newton says about rigid bodies (gravity, momentum, acceleration, and so on) is so well reflected by the observed behaviour of billiard balls that we quite easily forget about the abstraction of Newtonian mechanics and equate its propositions directly to the things we observe in everyday life. 'It makes sense,' we say, meaning that it is consistent with our senses.

If the laws of quantum mechanics are abstract, so are the laws of Newtonian physics. This relation between Newtonian mechanics and the illusion of everyday objects is exactly paralleled by the relation between quantum mechanics and the atomic system. Those who 'live with the atom', and have, by continued experimentation, become familiar with the 'behaviour' (i.e. the systematic results of measurements) of atomic systems, may learn to accept the statements of quantum mechanics with entire satisfaction, as providing a completely coherent and rational model embodying the main features of what has been consistently observed. What is missing, however, is the accompanying mental picture in terms of familiar objects. In hindsight, it is not as surprising as it seemed at the time of major atomic discoveries that the entities we classify as atoms, elementary particles, photons, etc., should behave not at all like billiard balls. It is the preconception that somehow they ought to do this that produces the main stumbling block for many who seek to grasp quantum mechanics. Basically, the 'model' provided by quantum mechanics is mathematical, and it can only be appreciated fully in mathematical language. Again, this is not really surprising because all the observations on atomic and subatomic systems are essentially quantitative in nature, and if they are to be related one to another, the tool for doing so has to be mathematical. It is helpful, however, even for the specialist, to have some sort of mental picture. And that brings us back to the 'waves' that have been discussed.

In classical physics waves are a subject of study distinct from mechanics. Clearly, some waves, like those on an ocean or in a vibrating crystal, can be interpreted as the combined motions of a large number of Newtonian particles. But in looking at an ocean-wave motion one ignores a great deal of possible information about the constituent particles in the wave.

It is no accident and not at all inappropriate that in popular language the same word, 'wave', is used, for instance, to describe the spread of an epidemic. We can think and talk of a 'wave of influenza' without putting names and faces to the individuals involved or without tracking down each individual virus particle or without following the movements of a particular human carrier who spread it. In other words, we have an epidemiological abstraction, which we can accept without demanding identification of all its items. In the same way, waves are a subject that mathematicians can study independently and that can be accepted (in some parts of physics) for the description of phenomena that are not reducible to the combined motion of many particles. For instance, a radio wave indisputably travels through intervening space since a receiver anywhere along the path intercepts it. But there is no need in the space for any particles whatsoever to make the transmission possible.

As has been pointed out, the abstractions forming the subject of Newton's light theories and Clerk Maxwell's theories are entirely different: separate 'paradigms'. What the founders of quantum mechanics found was that the behaviour of atomic and subatomic structures required a mathematical scheme that could accommodate both the *particle concept* as used in Newton's mechanics and the *wave concept* as used in Maxwell's theory. The entity we call an electron cannot be visualized by any picture of a billiard ball; what one needs is part of Newton's 'mass point' picture and part of the 'wave' picture. Since anything that is visible to us in everyday life cannot be conceived as any mixture of a particle and a wave, what has been said means that entities like electrons do not possess any close similarities with objects in our common experience. J. J. Thomson, who discovered the existence of electrons, without, of course, seeing them, said that he did not care whether they were pushed around by red-nosed pixies; what he was interested in was their behaviour.

Pixilated particles

Let us see, therefore, how pixilated physicists themselves are in their indirect observation of particles or light waves or whatever invisible matter they are studying. Suppose a metal is heated in a vacuum. Under suitable conditions it may be observed that a negative electric charge leaves the surface of the metal and moves in the vacuum (e.g. in a radio or television tube). This charge can be collected on other metal surfaces, or made to come into contact with all kinds of different pieces of apparatus. There are many ways of finding experimental evidence that both the emission of this charge from a heated surface and its acceptance by any piece of apparatus put in the vacuum container for that purpose does not proceed in a continuous smooth way. Empirically, the charge appears to be carried by a large number of separate parcels (consistent with Thomson's notion of electrons), each the same size as any other. Millikan showed that minute oil drops behave as if they carry just one, or two, or three, or more of these elementary units of charge, and he magnetically measured the unit. J. J. Thomson further noted that a beam of such units could be deflected in a magnetic field in such a way as would be predicted on grounds of Newtonian mechanics and the laws of magnetism if each charge were associated with a fixed unit of mass. The quantum physicist, without committing himself to a description of the 'parcel', accepts 'something' which he says has 'charge e and a mass m_e'. (This is like finding the fingerprints, but declining to draw an Identikit picture of their missing owner, in this case the electron.) The physicists in the early days of the electron's discovery were not so coy; they pictured the electron as a minute particle.

Nobel prizewinner J. J. Thomson's Nobel-prizewinning son, G. P. Thomson, was partly responsible for upsetting the picture of a pixie-pushed particle. In a series of experiments, streams of electrons were made to pass through an array of regularly spaced holes and their directions of flight after emergence from this lattice, or sieve, were inferred. ('Inferred' is an over-punctilious scientific word. Only indirect evidence in the form of spots on a photographic plate were seen, but since some agent apparently produced the spots, the posited electron's path could be 'inferred'.)

It was found that the electrons were emerging not in one beam and not in all directions at random. What did emerge were a number of separate, regularly spaced beams. What was seen was something well known in optics as an 'interference pattern'. Such a result was not predictable by classical *mechanics*. On the other hand, classical *optics* prominently features the notion that two or more waves can, according to circumstances, either reinforce each other or cancel each other out; so that, for instance, in passing through a grating or lattice a single beam of light splits into a set of regularly spaced beams, just like the beam of electrons.

In optics the existence of an interference pattern is understood to require that the several parts of a wave that take different paths through a lattice form part of the same general wave, matched in frequency and phase. By this principle, the pattern could never be produced by allowing pulses of light to pass at random, each through a different gap in the grating. This, in terms of G. P. Thomson's experiment, apparently rules out any possibility that one electron went through Hole One and another through Hole Two and came together again to produce the interference pattern. The inescapable conclusion had to be that each electron passed, in part, through *all* the gaps in the grating and, so to speak, interfered with itself. This indicated that any notion of localizing the electron had to be discarded. At the time, these experimental results were held to be paradoxical – for the very reason that localization was already part of the habit-formed mental picture.

But the apparent contradictions did not end there. Let us go back to the spots on the photographic plate. The picture of an electron emerging from the lattice as a spread-out series of beams and arriving at the detector as a single spot does not seem tenable. Each individual event is one of transfer of energy to the detector, say, the blackening of a photographic plate at a *single fairly localized spot*, and if the experiment is performed with very few electrons there will be only a few spots, but each of them localized. Agreed, these spots will only appear where they are allowed by the diffraction pattern. Where the diffraction beam is intense there will be many; where it is weak, few; and where it is cancelled out, no spots at all. The picture of a diffracted wave is palpably related to the beam of many electrons, but if we think of a single electron the findings do seem strange; in the end, how can it be that the electron, having spread itself out through all the holes in the grid, has arrived at one particular spot? If all but one of the grid holes had been closed there would not have been any diffraction pattern and the electron might have fallen on a spot that is 'forbidden' when the diffraction pattern is established.

It is worth noting that this creates the same paradox that arises when we consider light, although from the point of view of classical theory we have to start with an entirely different preconceived notion of light. Classically, light is a wave phenomenon, and diffraction and interference experiments are the best evidence of this. The detection of interference is by experiments exactly the same as those for electrons, say, by blackening photographic plates, and the actual effect observed on the photographic plate is a series of dots on each of which a 'packet' of light appears to have been concentrated so that it could react with one atom. There is nothing basically different between the black dots produced by electrons and those produced by light, and this led Einstein to his version of photons, already mentioned: particles of light that were quite unthinkable according to classical wave theory. In quantum mechanics, instead of having electron *particles* on the one hand and *light waves* on the other, each of these entities seems to combine the characteristics of particle and wave.

This concept of a 'wavicle' strains the ordinary imagination, but we might get some help from that eminent mathematician C. L. Dodgson, who in 1865 wrote *The Dynamics of a Particle* and, at the same time, in his more famous identity as Lewis Carroll, gave us *Alice* and the smile of the dematerialized Cheshire Cat. Or we might think of a cartoon in which the tracks of a skier are on a collision course with a tree. The left ski track goes round one side of the tree and the right ski track goes round the other but they merge again for the rest of the run. All one can assume from this evidence is that the body of the skier dematerialized on one side of the tree and rematerialized on the other.

The uncertainty principle

This either-or of quantum mechanics describes something, but does it explain anything? Physicists believe that it does in the sense that they have a precise mathematical model that incorporates features of the wave theory and other features of the particle theory that gives them the means of understanding in great detail a vast number of experimental results that would be contradictory in terms of either the particle theory or the wave theory. They would say that there are no paradoxes in the mathematics of quantum mechanics; paradoxes appear only when an attempt is made to limit explanation to the restrictions of everyday language. To illustrate, they could try to use the wave notion to the fullest possible extent and to ignore the particle notion as much as possible. They could get a fair way with this. For instance, they could pretend to forget the Bohr–Rutherford model of the atom, proposed ten years before Heisenberg, Schrödinger, and Dirac in 1925 contributed to evolve quantum mechanics. Bohr and Rutherford imagined an atomic nucleus surrounded by a group of electrons circling around as if they were planets moving round the Sun; this was a kind of solar system of particles. Modern physicists, however, describe detailed features of atomic, molecular, and crystal structure much more verifiably as a continual and continuous wave motion existing throughout the space surrounding the nuclei. They can give convincing descriptions of how the vibrations forming the waves are distributed in space, and the geometric distribution of these vibrations is the key to shapes and patterns that reveal themselves, for instance, in the detailed structure of crystals. But it does not explain what actually happens when two atoms hit each other and exchange energy and momentum. Instead of giving up the particle description altogether, they save it partially by saying that electrons are particles that are not governed mechanistically by the laws of cause and effect, as demanded by classical physics. Instead they are governed by *probability laws*.

These probability laws derive from the Uncertainty Principle formulated by the German physicist Werner Heisenberg in 1927. He showed that, in the light of this wave–particle duality of matter, precise determination of the position of a particle will be done at the expense of accuracy in specifying its velocity, and vice versa. He also showed that it would be logically impossible to determine exactly the energy of a system at any specific instant. The 'impossible' is emphatic; it is not a question of improving methods; any observational technique itself will impose the uncertainty. For example, supposing the velocity of a particle is known and it is decided to find its position by shining a light on it. The 'punch' packed by the photons of light impinging on the particle will change its momentum, invalidating the observed velocity. It seems to be an inescapable dilemma.

The uncertainty, however, applies rigorously only to an *individual* particle. Given a 'population' of particles, in which the position of one particle and the speed of other particles can be observed, it is possible to say with precision what the probable characteristics and behaviour of that *type* of particle will be. The quantum 'model' contains more than is involved in ordinary theories of probability such as are used for games of chance, population statistics, or even the description of complicated physical systems in statistical mechanics. Anyway, the 'exact' scientists find 'uncertainty' works.

The Heisenberg Uncertainty Principle appears to the majority of physical theorists to define with satisfactory precision just how far all possible observations can be related to visualizable models. Einstein, to the end of his days, was one of the doubters, unable to accept the probability description as something final.

And there are also those who would seek to prove that the particle aspect is basic and that the wave aspect might be explained in terms of a not fully understood field of force.

The Principle of Complementarity

Niels Bohr just refused to accept the 'either-or'. He argued that wave and particle were just two different ways of looking at the same thing; they were not mutually exclusive. He developed the Principle of Complementarity, according to which both interpretations are accepted as *complementary*, not *contradictory*, and are equally valid observations of the same phenomena. This prescription for peaceful coexistence, not of the protagonists but of the phenomena, need not be confined to subatomic physics. The neurologist, the biochemist, and the psychologist all study the brain somewhat differently, but their objective findings are not contradictory; they are complementary. The Principle of Complementarity, therefore, would appear to give a contemporary relevance to the theories of Anaxagoras. This Greek, of 2,500 years ago, without the benefit of $100,000,000 particle accelerators or quantum mechanics, had already arrived at some interesting notions. To recapitulate briefly: his universe, that is, the world and everything on it and around it, was composed of entities infinitesimal in size that were transmuted and re-embodied but eternally conserved. He sought to find a unifying principle that would account for the behaviour of each and every particle and each and every part of the universe and would govern all phenomena. He assumed the complementarity of the living and the non-living. He gave to his 'uniform substance' inherent, eternal, and organic properties. The extension of this to *nous* was consistent with the Principle of Complementarity – the regularity of the universe was rational, therefore *cosmos* (Order) demanded *nous* (Reason).

4 The Eternal Debate

Two thousand five hundred years from now students will look as patronizingly (or disparagingly) at present tenets of the cosmologists as we do at the picture that Anaxagoras had of the Cosmos. Professor Fred Hoyle, though satisfied that a thorough inquiry into the nature of the universe is not impossible, says:

> The astronomer is severely handicapped as compared with other scientists. He is forced into a comparatively passive role. He cannot invent his own experiments as the physicist, the chemist or the biologist can. He cannot travel about the universe examining the items that interest him. He cannot, for example, skin a star like an onion and see how it works inside.[1]

That will still hold after 2,500 years of space travel and the accumulation of all the information that it will yield and of all the supporting experimental work here on Earth or even in laboratories in space. H. G. Wells had one of his characters in the film *Things to Come* say as the manned rocket shot towards the Moon:

> For Man there is no rest and no ending – he must go on – conquest beyond conquest. This little planet, Earth, its winds and ways and all the laws of mind and matter that restrain him. Then the planets about him. And at last across immensity to the stars. And when he has conquered all the deeps of Space and all the mysteries of Time – still he will be beginning.[2]

In the never-ending debate on the nature of the universe, therefore, the farther we reach out, the more facts we acquire and confirm and the more experimental evidence we check and establish, the better informed the arguments become. In spite of the eminence of the authorities, however, the debate remains

1. Fred Hoyle, *The Nature of the Universe*, rev. ed., Harper & Brothers, New York, 1960, p. 3.
2. *Things to Come*, Cresset Press, 1935, p. 141.

open ended. The various contemporary versions of the universe are vigorously argued by their protagonists, who, it is assumed, have this much in common: each is fully aware of the latest information and verification. Nevertheless, now, as in the past and as it will be in the future, cosmology is a provisional interpretation of scientific facts. Some interpretations may be more persuasive and imaginative than others. Einstein said of a paper by one of his most distinguished colleagues: 'I enjoyed the data but I did not read the novel.'

Before we consider the various versions and the evidence adduced to sustain them, let us look at some notions that we accept as truths and that Anaxagoras did not know.

Anaxagoras conceived of the world as a circular raft, pneumatically supported like a kite. We know it as spherical; but, although he travelled and sailed over the horizons, he wrote as a prisoner of plane geometry. Nor did he, the 'inertial observer', seem to realize that his raft was rotating and spinning him around at a thousand miles an hour and that the raft itself was rushing through space at nineteen miles per second on its journey around the Sun. That, too, was inconceivable because, to him, the Sun was manifestly travelling around his anchored Earth, and man was the egocentre of the universe. But his universe was not just the firmament that he saw above him. The Sun and the stars went *under* the Earth and the Moon was eclipsed because the Earth screened the Sun's light from it.

In an age when mythology accounted for the Milky Way as a stream of milk, he described the galaxy as consisting of stars, just like other heavenly bodies that shine in the sky, but in greater density. He, however, apparently had no means of knowing, or suspecting, that in the Milky Way there could be at least a billion stars with planets, like our solar system, nor that it was only one of billions of such island universes in the Cosmos, as our optical telescopes and radio-telescopes testify today. Without a telescope he contemplated the Sun, the planets, and the stars and, remarkably, guessed the profound realization (2,500 years before the spectrographs confirmed it) that they were all formed of the same 'uniform substances'. He anticipated the atoms of Democritus; although his fundamental particles were held to be living (i.e.

organic matter), they were infinitesimal units, perpetually conserved but capable of combining and recombining in unlimited variety of forms from the seed to the Sun. These particles existed in *chaos* (Confusion) until churned into *cosmos* (Order) by *nous* (Reason).

That was a long time ago. We have observed and measured a great deal since and we are still trying, like Anaxagoras, to convince ourselves of the unity of Nature and to codify the 'laws' that determine it. But when we consider the massive dossier of evidence that modern science has assembled we should accept Einstein's reminder:

> The belief in an external world independent of the percipient subject is the foundation of all science. But since our sense perceptions inform us only indirectly of this external world, or Physical Reality, it is only by speculation that it can become comprehensible to us. From this it follows that our conceptions of Physical Reality can never be definite; we must always be ready to alter them, to alter, that is, the axiomatic basis of physics, in order to take account of facts of perception with the greatest possible completeness. A glance at the development of physics shows that this basis has in fact suffered profound modifications in the course of time.

There should be no night

There is one statement in that version of cosmogony, the Biblical book of Genesis, with which modern scientists (while reserving their ideas about the deity) would not disagree: 'God said, "Let there be light" and there was light! ... and God divided the light from the darkness.' What is known about the universe of matter derives in large measure from the nature and behaviour of light. If night were not divided from day, by the obscuration of the Sun, we could not see the stars, millions of which are intrinsically far, far brighter than the Sun. And the curious phenomenon of darkness provides Olbers's Paradox of 1826 which is one of the bases of modern cosmology.

The paradox is simple, and yet profoundly significant. Olbers observed, as any of us can do any clear night, that there are some very bright stars, some not so bright, and some very dim. The easiest assumption to make is that the brightest stars are the nearest, the medium-bright stars are farther away, and the dim ones farther away still. We can also intelligently assume that there are stars even more remote that are so faint that they cannot be seen individually. But Olbers asked whether those remote stars might not be so vastly numerous as to floodlight the night sky. He found that in trying to find effects from regions too far away to be seen in detail he had to make assumptions about the depths of the universe. And the assumptions he made are reasonable in terms of what was learned after his day. He first assumed that the distant regions of the universe would be consistently like our own immediate celestial neighbourhood. He estimated that there would be stars there with similar radiation and spaced at the same average distance as the stars we can see. In other words, he assumed a symmetrical dispersal in all regions of the universe. He accepted as fact that the laws of physics would apply in the uttermost regions as well as in the innermost: for instance, that light spreads out after leaving its source, as the light spreads from a candle. Lastly, he made the assumption that the universe was static, so that the fixed stars would be continuously propagating light from their prescribed positions in the heavens. Thus the universe would consist of a series of layers of stars (like a luminescent onion). If concentric layers are added, one to another without limit, the amount of light the Earth would receive from each layer would be the same, regardless of the radius of each layer, and the amount would increase so that an infinite amount of light would be received from layers stretching to infinity. There is an obvious limiting factor because there would be shadows arising from intervening stars that obscure those beyond. This would keep the sum of light from adding up to infinity. Nevertheless, the sum, reduced by such subtractions, still gave Olbers an incandescent glow theoretically equal to 50,000 times the light from the Sun at its zenith. He tried to account for the common observation that this is not actually so by assuming that the reduction was due to obscuring clouds of non-radiating matter in space. This did not help him because this matter would be absorbing heat until it, too, would radiate secondhand heat and would no longer serve as a screen. Many attempts were made by Olbers and others to escape from this dilemma. His physics, based on the assumption of fixed stars, was inescapable, but his sum would have meant that not only would there be no night and day but that we would have been subjected to temperatures of 10,000°F. (5,536°C.), whereas the highest earthly temperature is around 120°F. (49°C.).

Olbers's Paradox helped turn cosmology into a science because it made men of science re-examine his assumptions and test them by observation. For one thing, we observe that the stars are not disposed like the spectators in an infinite sports arena. They form galaxies – a series of stellar arenas – and we perceive, too, that there are billions of other galaxies like our own Milky Way. This would not have invalidated Olbers's assumptions because for 'stars' read 'galaxies' and his argument would still hold; the galaxies and the clusters of galaxies would still have given him his incandescence. Nor does his assumption of the here-and-now of light (whereas we find that light travels with a speed of 186,300 miles per second and that the light he was observing from any star had originated, not on the night he was looking, but millions of years before).

No, the root of Olbers's Paradox was his assumption of a static universe. Perversely, his theories demonstrated that the universe *must* be expanding. Before we even consider the why and the wherefore and the whither of the expanding universe, in our contemporary debate, the fact that the night sky is dark and that we are not incinerated by Olbers's 10,000°F. are evidence enough. Why? Because, if the universe is expanding, then the distant stars are moving away from us at highest speeds and a simple truth of physics is that that light emitted by a receding source is reduced in intensity compared with the light emitted from a source at rest. The clusters, the galaxies, or the individual stars are escaping as fast as they can from Olbers's Paradox.

But the present evidence for the expanding universe lies in the highly refined methods and measurements used in studying what Francis Bacon called 'God's first creature, which was light'. And we can legitimately define 'light' as 'all electromagnetic radiation', so that we can include the new evidence of radio-astronomy.

The methods and measurements are provided by spectroscopy. The simplest example of a spectrum, which the spectroscopists examine, is the rainbow. This is simply the breaking up of the white light of the Sun into its component colours (or component wavelengths). A glass prism, or a grid system, can do the job more precisely. The sunlight can be spread out, like a scroll, and the colours separated into their wavelengths: blue on the left and red on the right. But the colours do not form a smooth, continuous band. In places, dark lines run across the spectrum. They are mainly due to the light of the Sun shining through the cooler gases of the Sun's atmosphere and those gases happen to be opaque to very particular colours and produce the effect of thin shadow lines. Modern astronomers have powerful telescopes and spectroscopes of high susceptibility and spectral precision so that they can analyse the light of individual galaxies and individual stars. The very distant galaxies yield very little light and therefore the spectrum will not be as clear as the spectrum of the Sun, but they do show the prominent dark lines. And these lines are not where they would be in the matching spectrum of the Sun. They are 'shifted'. The shift is always towards the red, that is, to the right. The fainter the galaxy looks the greater the shift to the red.

The red shift

In wave physics the shift towards the red end of the spectrum always indicates the velocity of recession. The 'Why?' is more easily understood in terms of that other wave phenomenon: sound. You are standing in a railway station; an express train is approaching at high speed, with its whistle blowing. As it passes you, the pitch of the whistle drops suddenly. This has nothing to do with the whistle nor with your ear; it is due to the speed of the train. And for the following reason: suppose the velocity of sound to be exactly 370 yards per second. The approaching whistle is 370 yards away; half a second later it is 185 yards away. Treat them as two separate instants and it will be seen that the second sound reaches you, the stationary listener, only half a second later than the first. Now, instead of considering them one second apart, let us say that the high-pressure points of the sound wave are a thousand cycles, so that the peaks of the wave are only a thousandth of a second apart. In that thousandth of a second the train whistle will have moved ·185 of a yard and the sound after that will take only two-thousandths of a second to reach you on the station platform. The wavelengths are being shortened owing to the movement of the source, and being compressed to a higher pitch, or shrillness. As the engine passes you the process is reversed and each successive sound wave has a greater distance to travel and so the interval between the successive peaks is longer than the interval between their emission and accordingly the pitch of the sound will be lowered. This is what is known as the Doppler Effect.

The same thing applies to light waves, but in their case the speed is not that of sound, about 1,100 feet per second, but of light, which is 186,300 miles per second. The instrument is not the human ear but the spectroscope, spreading out the radiations into their colour frequencies. The visual equivalent of an increase in auditory pitch is a shift of the spectral lines towards the violet end of the spectrum, and a decrease of pitch, i.e. an increase of wavelength, is equivalent to a shift towards the red end.

A 'red shift', therefore, indicates the velocity of recession of the source (e.g. a star). The velocity of recession stands to the velocity of light in the ratio given by the magnitude of the red shift: the change in the wavelength divided by the wavelength. If the faintness of a galaxy is accepted as an indication of its remoteness and the red shift of the spectra as the velocity of recession, then the velocity of recession is proportional to the distance of the object.

This, as will be seen, is the transfer to remote space of physical experience observed and measured on Earth. It is valid only if there is universal uniformity: that there is really a *cosmos*, i.e. Order. This has to be assumed before it can be examined. And, the more efficient our means of observing – more powerful optical devices and radio-telescopes – the more convincing the arguments of uniformity become. If the medley of the heavens is sorted out and we look close enough (in detail) and far enough (in distance), remarkably consistent patterns emerge and repeat themselves. It is like the design of a wallpaper; the details may vary, but the pattern is reproduced consistently and, as you look around the room, is the same in all directions: the ceiling pattern is the same and it is repeated in the carpet.

If, by rigorous and repeated observations, one is satisfied that there is a general pattern, and when each new discovery is consistent as a detail in that pattern, then one is entitled to assume that this is one of the uniformities of Nature (a description which J. B. S. Haldane insisted was more proper than the term 'law'). It is also permissible to assume that that pattern will appear the same from every point of vantage whether it be that of an invalid lying in bed studying the cosmic wallpaper, an intelligent fly on one of the galactic designs on that wallpaper, an observer at Mount Palomar looking at the universe through a powerful telescope, or his 'opposite number' on a planetary world in another island universe looking through a similar telescope.

Pushing back the galaxies

In 1923 Edwin P. Hubble looked at this pattern in a particular way. At Mount Wilson Observatory he was studying the Spiral Nebula of Andromeda. A spiral nebula is a characteristic feature of the repetitive pattern; it has a centre shaped like a lozenge with spiral arms winding round it. Until Hubble had another look at Andromeda, these nebulae were supposed to be located among the stars of the Milky Way. He noticed that the spiral arms of Andromeda contain a number of extremely faint stars, the brightness of which changed periodically. There is a generic term for such stars: 'Cepheid variables'. Their periodic changes in luminosity are explained by the periodic pulsations of their giant bodies. The brighter the star, it has been observed, the longer the period of pulsation. The Harvard astronomer Harlow Shapley had used those pulsations as a means of measuring distance; by measuring the pulsations he could establish a star's absolute brightness and this, in measured contrast to its visible brightness, gave him the actual distance to such a star. Using this method, Hubble assured himself that those stars in Andromeda had high absolute brightness and this, in measured contrast to its visible brightness, gave him the actual distance to such a star. Using this method, Hubble assured himself that those stars in Andromeda had high absolute, initial luminosity and yet were so faint visually as to be at the limit of visibility. (This is like saying that a powerful searchlight showed up as feebly as a match.) If this were so, then there was only one possible explanation: that they must be very, very far away. By systematic measurements of this kind it was shown that Andromeda, so far from being a lodger in the Milky Way, was a galaxy in its own right, at least a million light-years away. Walter Baade, also of Mount Wilson Observatory, was able to resolve photographically the central body of Andromeda and the associated swirls and to sort out individual stars. It is now known that Andromeda and many, many more spiral nebulae each consists of billions of stars like, or greater than, our own Sun. What had once looked like luminous sandstorms have emerged as island universes. The individual stars blur together into a faintly glowing mass only because of their distance away from the observer.

Hubble's discovery of new tracts and vast distances in the universe was followed up by the spectroscopists. The light emitted from the spiral nebulae showed a 'shift to the red'; that meant, in terms of the Doppler Effect, that they were moving outward, ever outward, stellar wagon trains pushing out to the new frontiers. In the diorama of the universe, those nebulae which had once seemed part of the foreground of the pattern of the Milky Way are now seen to be at remote depths. The scientists now recognize that the entire space of the universe is populated by billions of galaxies – not only the upgraded 'spiral nebulae', but elliptical and spherical galaxies – each with billions of stars with their own solar systems of planets. What is more, all the galaxies and the stars composing them are flying away from each other at fantastic speeds.

The notion of an expanding universe is physically and mathematically unavoidable if we are to escape from Olbers's Paradox (and incineration by the incandescent fixed stars) and to account for the red shift – velocity – distance. To express it as a mathematician would: the universe has uniformity. How can it move and still maintain its uniformity? The answer is that it can only move in such a way that the velocity of every object is in the line of sight and proportional to its distance. This is the only type of motion which will maintain uniformity. Therefore, an expansion with the velocity of recession proportional to distance is a natural consequence of the assumption of uniformity which is based on confirmed observation. Any theory of the universe consistent with the observed facts must always come up with the answer that it must be in motion with objects showing velocities proportional to their distances.[3]

Taking further liberties with the pattern of the universe, let us, instead of draping it on the walls, put it on a child's balloon. Just as the Declaration of Independence can be photographed down to the size of a printer's full point or the 'microdot' of the spy thrillers, so we can print on the balloon all the clusters, galaxies, and stars. (Our Sun would be just an atom of carbon in the printer's ink.) As we proceed to blow up the balloon the congestion begins to sort itself out and each item in the design moves farther and farther away from the other items. The galaxies become differentiated in the clusters, and the stars in the galaxies become separate specks. What this model shows is the recession due to the uniform expansion of the entire system; what it cannot show is that galaxies also possess an individual random motion similar to the thermal motion of molecules in a gas, and the two kinds of motion interact and sometimes, if the random motion is towards the observer, the spectrum may show a violet shift (to the left) instead of a red shift (to the right). This might seem to belie the general argument of expansion, but does not, because at greater distances the increasing recession velocities are too great to be affected by the random 'thermal' velocities. Thus the expansion of the system as a whole is not challenged.

The Expanding Universe

The Expanding Universe is the reigning paradigm. Astronomical observations provide more and more supporting evidence and the scientific consensus accepts it. Anyone who disagreed would

3. Cf. Hermann Bondi's *The Universe at Large*, Doubleday Anchor, Garden City, New York, 1960; Heinemann Educational Books, 1962.

be considered as eccentric as a Flat-Earther. The cosmological debates in learned societies, which sometimes have the gusto of a bar-room argument or a meeting of the Pickwick Club ('alleged theory', 'so-called astronomer', or 'Marxist interpretation'), are not about whether it is expanding, but 'Why?' or 'How?' or 'In what direction?' Because while Einstein's curvature of space is accepted, is it spherical (roughly like our globe) or is it hyperbolic (like a saddle)?

In fairness to the protagonists, and the vigour of their arguments, it should be recognized that none of them would claim for a preferred theory that it is true, much less infallible; it is not fixed, but fertile. A theory has to have the substance of evidence and has to link and carefully correlate observations with any mathematical model proposed. Any theory is an Aunt Sally to be shied at by other scientists, and no reputable scientist, however devoted to his favourite Aunt Sally, would ever withhold ammunition or ignore a direct hit by another theorist. The purpose is not to stun the opposition but to stimulate, and predict, new observations.

The cosmological theorist says to the practising astronomer, 'Look out for so-and-so,' and starts a new chain of inquiry. Or, as has often happened, the radio-astronomer will say to the optical astronomer, 'I am getting powerful radio signals from a source which does not appear on your star charts. See if your telescope can locate it.' For example, two of the most powerful radio sources are on the bearings of the constellations Cassiopeia and Cygnus. At Mount Palomar Walter Baade used the telescope to examine the Cygnus region and found that the radio source was visible. It was 270 million light-years away. The distance and the power of the signals showed that this was indeed a remarkable phenomenon. Various explanations have been offered for 'Cygnus A', this transmitter operating with a strength of 10,000,000,000,000,000,000,000,000,000,000,000 kilowatts. One is that a star (a supernova) exploded and triggered off a chain reaction of other star explosions, causing the production of electrons travelling with nearly the speed of light and causing intense radio waves. Another (now generally discarded) was that it was a collision of two galaxies. Another is that so far from being a collision or an explosion, it is the birth of a galaxy. In this process the gas cloud (the proto-galaxy) contracts and breaks up into smaller clouds which become more dense and form the stars. As this happens, gravitational energy is released which generates cosmic rays. These collide with gas atoms and discharge great amounts of high-speed electrons necessary for intense radio emission. Thus there are three (and there may be more) different theories of a phenomenon, indicated by the radio-astronomers and observed by Baade. None of them is 'true' and none of them is 'false'; each is waiting for evidence, being vigorously sought all the time, that will support one against the other, or discard the lot.

How such evidence is arrived at and applied is exemplified in two 'tidy' examples. In the year A.D. 1054 the Chinese witnessed and recorded a celestial catastrophe – a star exploded. There are obscure stars which can suddenly blaze up, with a burst of light millions of times the luminosity of the Sun, and then they dim out in days, or months, or years: a wink of an eye in the chronology of the Cosmos. They are called dwarf novae, novae, or supernovae. Dwarf novae are minor affairs: *merely* equal to 20 to 200 million million hydrogen bombs or the equivalent of 100 million tons of T.N.T. Novae multiply that by another million. Supernovae are 100 million times brighter than our Sun. Dwarf novae and novae may repeat the performance at intervals, years apart, like a dimmed lamp being switched up. But a supernova is a once-for-all; in its explosion it releases its energy. It disintegrates into gases.

When Einstein's General Theory of Relativity led him to the concept of curved space, he had to take a hard look at the universe. The notion that it was an infinite expanse in which galaxies and stars were floating like lilies in a pond had to be abandoned. His reasoning at first led him to a universe which curved back on itself and closed up, like a globe. To test this idea he assumed a simplified model in which matter is not present in the shape of isolated lumps but rather as a thin fog spread uniformly and filling all space. To simplify his thinking still further, he assumed that the universe was static so that the density of matter would always remain the same and the distances between points in space did not undergo any kind of systematic change. This led him into difficulties with his own theory; he had to alter his fundamental gravitational equations by adding another term which was the equivalent of assuming that a small negative pressure existed everywhere in space.

The Dutch astronomer De Sitter examined Einstein's version. He found that Einstein's universe would not hold together. If matter were imprisoned in a cosmic jail, there would be an inevitable jailbreak due to the new factor which Einstein had introduced: negative pressure, which would manifest itself as repulsion. Therefore De Sitter argued that the distant objects would behave as though they were receding from us. He suggested that astronomers ought to study the light from the distant nebulae to see whether there was a shift towards the red end of the spectrum.

In the next five years, forty spiral nebulae were systematically examined. There was a measurable red shift in all but four, and the exceptions were the nearest neighbours of the Milky Way and, predictably, from the Doppler Effect (velocity–distance), would not give what might be called the 'weary wave' drag to the right of the spectrum.

Hubble, using the hundred-inch telescope at Mount Wilson Observatory, extended the examination and made the discovery that not only are the distant spiral nebulae receding but that the speed of recession increases with the distance. He was the first to state categorically that the universe was expanding.

That was on the strength of direct observation, but at the same time the mathematicians were arriving at the same conclusion. In 1922 the Russian mathematician A. Friedmann re-examined Einstein's model of the universe and found that he could explain the observed recession without introducing a negative pressure, or cosmic repulsion, term. He kept the original general relativity equations unchanged, but allowed the curvature of space to change with time. Abbé Lemaître in Belgium, Robertson and Tolman in the United States, and Sir Arthur Eddington in Britain were all independently producing theories and all of them agreed that the universe was expanding.

The Big Bang

Expanding from what? The point is that the galaxies are not only receding from us, the earthly observers, like runners in a cross-country race; they are scattering, running away from each other and from every point in space. If the distant galaxies are thus receding and diverging, obviously they must have been much closer together and, if one goes much further back in time, may have formed a supercondensed state of matter.

This brings us to the Big Bang concept, which is logically basic to the several relativistic theories of the origin of the universe. A simple way to look at it is as a nuclear bomb explosion, of which we make a continuous film record. We see it go off and we see it expand into the mushroom cloud of particles, escaping from each other in ever-widening dispersal. Then we reverse the film, the way they do in trick movies when they put the high diver back on the spring-board, and we see the mushroom cloud reducing back into the dimensions of the bomb and, if we could follow, into the nuclei of the atoms. In the same way, we can take the evidence which astronomers have accumulated of the distance between the galaxies doubling every 2,600,000,000 years and imagine the reverse. It is like putting an Arabian Nights genie back in the bottle. The size and nature of the original 'bottle', however, is a matter for debate.

According to the Lemaître picture, the universe started as a giant, primeval atom which exploded many billions of years ago. As the hot gases cooled they condensed into galaxies, stars, and planets. In Eddington's picture, our present universe is a compromise between the static universe which Einstein suggested (and then abandoned) and the empty, expanding De Sitter universe. According to Eddington, the universe which was the precursor to ours was a relatively small static sphere having a radius of 1,068,000,000 light-years. Somehow at this stage it was triggered and began expanding and would go on expanding until the galaxies were infinitely dispersed. He was led, by what his contemporary Sir Oliver Lodge referred to as 'My friend Arthur's little sums', to a universe which is curved back upon itself (elliptical, not hyperbolic, or saddle-shaped). The horizon of a spherical universe, however, would not be an ultimate limiting factor, since light, in space–time terms, is also curved, and we could, in effect, see the back of our cosmic sphere. The limiting factor would be the velocity of light: the speed limit of the universe. 'Light', said Eddington, 'is like a runner on an expanding track with the winning post receding faster than he can run.' At a distance of 12,000,000,000 light-years the galaxies would be rushing away from each other at 186,300 miles an hour, the ultimate speed. Light receding from us at that speed and that distance would never reach us. We would not be able therefore to observe by optical astronomy or to listen by radio-astronomy, since the speed of light is the speed of radio as well; but that will not limit further speculation, because we can study the other runners in the relay race of the galaxies, and judge how they are speeding up.

The expansion of the universe can take place in three different ways: it may be expanding at a constant rate; or the expansion may be speeding up; or the expansion may be slowing down.

The pulse of the universe

In 1965 observations and confirmation of previous observations introduced interesting new factors into the arguments. Dr Allan R. Sandage of Mount Wilson and Mount Palomar observatories reported the discovery of numerous faint blue objects which he called 'quasi-stellar blue galaxies' (Q.S.B.G.). They had been observed during the previous twenty years, but regarded only as local blue stars in the outer regions of the Milky Way. They did not emit radio signals. Around the same time, Dr Maarten Schmidt also announced his analysis of light coming from five other extremely remote, radio-emitting objects, quasi-stellar radio sources (quasars). Until 1963 they too had been regarded as normal, but faint, stars in the Milky Way.

In the diorama of the universe, in the stereoscopic depths of space, those stellar objects, previously regarded as details – sequins – on the Milky Way drapes of our proscenium, had now been revealed as at great distances, at the back of the setting. The light received from them, with or without radio signals, showed very large red shifts.

(In case there is pardonable confusion about a 'blue galaxy' having a 'red shift', one has to remember that the red shift has nothing to do with observed colour; it means that the lines on the spectrographic pictures are farther to the right – towards the red end of the spectrum – than normal.)

The shift was so large as to give a new extension to the universe. If celestial objects are plotted on a graph of which the vertical axis is the red shift (related to increasing speed) and the horizontal axis is decreasing brightness (related to increasing distance), then the plot would give an indication of the nature of the expansion of the universe. A straight line would indicate an open, infinite universe, with the expansion continuing constant. If, in the outreaches of the observed universe, the plot curves towards the right, it would show an open, infinite universe with the expansion speeding up. But if the plot curves towards the left, turning in on itself, it would show a closed, finite universe with variations and pulsations, and indicate that the expansion is slowing down. Significantly, the quasi-stellar blue galaxies and the quasi-stellar radio sources were, on the graph, in the distant reaches, all to the left of the straight line. Sandage proposed (one should not say 'concluded') from those first findings that the universe is finite, sphere-shaped, and slowing down in its expansion.

It follows that, if the expansion is in fact slowing down, it will stop and the process will reverse itself and the universe will contract. We have here the notion of a pulsating universe, which is still quite consistent with the Big Bang and the Expanding Universe. It just raises questions of 'eternity' in which the Big Squeeze (a capsulized universe) explodes as the Big Bang, with rapid expansion, followed by a contraction into another Big Squeeze, followed by another Big Bang, and so on. Sandage on his evidence was able, in 1965, to work out the 'pulses'. He estimated the pulse interval, from Big Bang to Big Bang, as about 80,000,000,000 years – 40,000,000,000 years outward expansion and the same period for contraction. If we look at Lemaître's original suggestion of a compressed primeval atom, or Eddington's considerably larger concentration, they could fit into the argument as versions of the Big Squeeze which was followed by the Big Bang which gave the *present* version of the universe, without prejudice to previous, or future, ones.

George Gamow, in his lusty support of the Big Bang, has been more explicit. In his *The Creation of the Universe* he anticipated the quasar/blue galaxy evidence. He described the high state of compression which preceded expansion. He estimated that all the matter (galaxies, stars, planets, etc.) within reach of the twentieth-century Great Eye of Palomar (the 200-inch telescope) was encompassed in a sphere only thirty times as large as our Sun. Each cubic centimetre of space, in this concentration, would have contained 100 million tons of matter!

> ... why was our universe in such a highly compressed state, and why did it start expanding? The simplest, and mathematically most consistent, way of answering those questions would be to say that *the Big Squeeze which took place in the early history of our universe was the result of a collapse which took place at a still earlier era, and that the present expansion is simply an 'elastic' rebound which started as soon as the maximum permissible squeezing density was reached.*[4]

Gamow held that the density reached at the maximum of compression must have been so high that any structural features which may have existed in the pre-collapse era were completely obliterated and that even the atoms and their nuclei were broken up into the elementary particles (protons, neutrons, and electrons) from which they were built. The universe would start all over again, from scratch.

Continuous Creation

Protagonists of the Big Bang theory might spar among themselves over the various versions, but they were in head-on collision with the school of thought which sustained the Steady State, or Continuous Creation, theory. This is inconsistent with either a once-for-all explosion or the repeat performances of the pulsating version.

4. Gamow, op. cit., p. 29.

Fred Hoyle, Plumian Professor of Theoretical Astronomy and Experimental Philosophy at Cambridge University, while accepting an *expandable* universe, did not like the idea of an *expendable* universe. He accepted the evidence that every galaxy we now observe is receding from us and will, in about 100,000,000,000 years, pass entirely beyond the limit of any observer with any conceivable astronomical telescope, based in the world or in our own galaxy. This, if the Big Bang theory were to prevail, would produce empty space, apart from one or two of the very near galaxies which the Milky Way would retain as satellites. The billions of galaxies, with their billions of stars, would have disappeared.

Hoyle advanced his own theory of Continuous Creation, which would ensure that astronomical posterity, even 100,000,000,000 years from now, would see about the same number of galaxies as we see today.

> New galaxies will have condensed out of the background material at just about the rate necessary to compensate for those that are being lost as a consequence of their passing beyond our observable universe. At first sight it might be thought that this could not go on indefinitely because the material forming the background would ultimately become exhausted. The reason why this is not so is that new material appears in space to compensate for the background material that is constantly being condensed into galaxies. . . . I find myself forced to assume that the nature of the universe requires continuous creation – the perpetual bringing into being of new background material.[5]

Independently, two other eminent British professors, H. Bondi and T. Gold, while using a different form of argument from Hoyle, arrived at a similar conclusion. Continuous Creation can be represented by mathematical equations, the consequences of which can be worked out and compared with observation.

The basis of the Steady State cosmology was the assumption that the universe was not only uniform in space but also unchanging in time, when viewed on a sufficiently large scale. Owing to the expansion of the universe (accepted), the mean density of matter would appear to be diminishing all the time, contrary to the assumption that the system is unchanging. To avoid this con-

5. Hoyle, op. cit., p. 102.

tradiction the 'Continuous Creators' suggested that there was a sort of cosmic supermarket in which goods were being continually removed from the shelves, to be replaced from the storeroom which, in turn, was replenished from the manufactory.

The source material of the stuff of the Cosmos is hydrogen. Inside the stars (as in our own Sun) hydrogen is being steadily converted into helium. Throughout the universe this is a one-way process. Only to an insignificant extent is hydrogen produced by the breakdown of other elements. Yet the universe consists preponderantly of hydrogen, the simplest element of all (one proton and one electron) and, the Hoyle–Bondi–Gold argument ran, this would be impossible; the hydrogen supply would become exhausted. Therefore, atoms of hydrogen must be coming into existence all the time. Hoyle saw no difficulty in this. Matter gives rise to a gravitational field. Now he suggested we must think of it as giving rise to a 'creation field'. Matter begets matter.

Bondi compared the Steady State with the human population. Each individual galaxy ages, owing to the way its resources of hydrogen are being depleted by its conversion to helium inside the stars. However, the ageing of the individual members of the universe does not imply that the universe as a whole is ageing. Like the human population, each individual object of the universe is born, grows up, grows old, and passes on (into the Far Yonder, whose bourne is the speed of light). But assuming that new matter is being conceived, the general picture does not change at all – if the system is uniform and the disappearing galaxies are compensated for by an equal amount of newly created matter.

The rate of making 'atoms of hydrogen out of nothing' (Bondi's phrase) is very low indeed owing to the tenuousness of the distribution of matter, which means that coincidence in the 'creation field' is rare – like the chances of Java Man meeting Peking Woman in the geography and time scale of human evolution. In relation to the whole volume of the Earth it would amount only to a mass like that of a particle of dust every million years.

Umpiring the bout

There are, however, tests which can be applied to the contending cosmological theories. As has already been made clear, if we look at the distant regions of the universe we do not see them as they are now. The light (or radio waves) which we receive was sent out billions of years ago, identifying the objects as they were those billions of years ago. According to Lemaître, and his 'hypothesis of the primeval atom':

> The atom world broke up into fragments, each fragment into still smaller pieces. Assuming, for the sake of simplicity, that this fragmentation occurred in equal pieces, we find that 260 successive fragmentations were needed in order to reach the present pulverization of matter into our poor little atoms which are almost too small to be broken further. The evolution of the world can be compared to a display of fireworks that has just ended: some few red wisps, ashes and smoke. Standing on a cooled cinder, we see the slow fading of the suns and we try to recall the vanished brilliance of the origins of worlds.[6]

The hot ashes which are left are the galaxies. (Cosmologists can afford to be poetic; there is no time limit on their licence.) When the 'firework' went off (say, 40,000,000,000 years ago) and released a great density of matter, the rate of the expansion produced by the explosion was slowed down by a strong force of gravitation.

In the general theory of relativity, there is also the universal force of repulsion, increasing with distance. As long as there was a relatively compact mass of matter, gravitation was much more powerful than this long-range force of repulsion, but as the system expanded it approached a state in which the force of gravitation exactly balanced the force of repulsion. This should have produced equilibrium, but the brakes of gravitation kept slipping and the components of the universe went freewheeling faster outwards.

According to Gamow, after the full complement of atoms had been formed during the first hour of expansion 'nothing of particular interest happened for the next 30 million years'.[7] As the hot gases continued to expand, and to cool off by expansion,

6. Gamow, op. cit., p. 53. 7. ibid., p. 74.

the atoms congealed into dusty particles and the gas-particle mixture condensed into clusters of galaxies and these again into individual galaxies.

Comparing Anaxagoras's theory (the celestial butter churn) with that of Lemaître (the cosmic firecracker) they both agree that once Cosmos emerged from Chaos, that was that: the pattern was fixed, with Lemaître's provision that it was expanding. On this basis, all the galaxies have the same birth certificates. So if we study the nearer galaxies, they should tell us all we want to know about the farther galaxies. We could *assume* that the pictures would be the same and the designs – spiral, elliptical, or spherical – would be repeated. The fact that the pictures we actually see are a bit faded does not mean that the objects are older than the ones we see nearer at hand. On the contrary, when they posed for those pictures, the galaxies were younger by the billions of years that the telephoto has taken to get here. (Bondi provided the neat analogy of twins, one in England and the other sent to live in Australia. If pictures are regularly sent home by the Australian twin he will always look younger by the weeks taken by surface mail.) But since we have no way of knowing, so far, what a young galaxy looks like compared with an old one, comparative observations do not help.

In the Steady State theory the circumstances would be different. The galaxies were not born of a single litter; they are being born all the time. As the universe expands, the older galaxies, 'move over' to make room for new ones. But in the distant regions as in the nearer, the *average* age distribution is the same; none of the *average* features would change with distance. One of the tests which the practising astronomers can apply to the theory is to see whether any features of the galaxies vary with distance.

Another test is the number of galaxies of more than a given brightness. As has been pointed out, a searchlight at a great distance may look like a match; the dimmer a star appears, the presumption is that it is far away. There should be a greater concentration of faint, distant galaxies according to the Big Bang theory than according to the Steady State theory. Why? Because, due to the billions of years that it has taken the light to reach us, we are looking at a billions-of-years-old picture and expansion has been going on in the meantime. Although it will be billions more light-years before we *see* those distant 'nursery' galaxies, the radio-astronomers might even now be able to *hear* them in their prenatal state, or pick up their birth cries. They have a guide: galaxies have a certain size (about 80,000 light-years in diameter) and, according to Steady State, should have a certain average distance apart (about one-and-a-half million light-years).

The Steady State theory assumed eternity, a universe without beginning and without end. It is like a continuous cinema performance that goes on reel after reel, forever. But, as we did before, let us run the film backwards. The Expanding universe would become the Contracting universe. All the galaxies (including those which, in the forward run, Hoyle, Bondi, and Gold had newly made for us) would begin to close in. And in the reverse process they would dematerialize. Each star in the galaxies would disintegrate into the gases from which they were formed and those proto-galaxies would relapse into the incoherent state from which they condensed: the spontaneously created atoms. Before the neighbouring galaxies converged on our Milky Way, threatening to crush it, they would fade out into tenuous space. But that would not save us because, after we had run back through 15,000,000,000 years, our Milky Way would have vanished too, and long before that the Earth would have disappeared. But the film, backward-run, or forward-run, goes on forever.

This gets us out of a lot of philosophical difficulties – like St Augustine of Hippo's question 'What was God doing before he made heaven and earth?' or the origin of Anaxagoras's *nous* – but it still does not suggest the origins of Anaxagoras's organic elementary particles, nor the physical beginning of the nuclear particles, nor the primeval hydrogen. What we have is the 'atoms out of nothing' of the Steady State theory.

What distinguished the Steady State, or Continuous Evolution, theory was that it could be tested fairly simply. It predicted that the spatial densities of galaxies should not change with time. On the basis of the other theories, the spatial density would have been much greater in the past than now. With optical telescopes, the range is equivalent to looking back 2,000 million years and, within that range of space and time, there was not any change in the overall uniformity. That reinforced the Steady State theory. But the radio-astronomers reached out far beyond the optical limits. Their mapping of the universe listed nearly 10,000 radio sources and, as their instruments and methods became more sensitive and discriminating, the results indicated that the spatial density of radio sources increased as their intensity decreased. On any logical interpretation this implied an increasing density with distance, and that would be inconsistent with Steady State. At first the findings were disputed, but when Sir Martin Ryle at Cambridge, England, refined his techniques and increased the number of counts, Hoyle accepted the results of the 'poll' and conceded the defeat of the Steady State theory in its original form. As he remarked wryly, 'The present day astronomer is just as powerless to predict future developments as was his predecessor of thirty years ago.'[8]

The quasars were an exciting new factor, but they complicated rather than simplified thinking about the nature of the universe. If the red shift really means that the sources are receding, then some of them are moving away with a velocity 80 per cent that of light, which means that they are more than 8,000,000,000 light-years away and that they represent the early processes in the evolution of the universe. There are, however, doubts as to whether they in fact represent recession in distance. If the quasars are at immediate distance, i.e. in, or around, nearer galaxies, the red shift could be explained by explosions in those near-by galaxies. The energies involved are prodigious and theories have suggested a gravitational collapse of a concentrate of 100 million stars and even the annihilation of matter by anti-matter. What this means is that the new discoveries and observations have raised more questions than they have answered. Present measurements cannot determine whether the universe was in a compressed condition 10,000 million years ago and exploded (Gamow); whether it was in the primeval atom condition 50,000 million years ago (Lemaître); or whether it pulsates, i.e. expands and then contracts to a state of high density. One circumstance favours the latter. That is the background microwave radiation observed by Professor Robert H. Dicke, of Princeton. He claimed (1966) to have received

8. *Science Journal*, vol. 2, no. 10, October 1966, p. 3.

electromagnetic waves consistent with emissions generated in a cosmic fireball of fantastic temperature. These emissions he had identified as the microwaves which would accompany the creation of helium 200 seconds after the original explosion and this he puts forward as proof of the pulsating universe.

In weighing the arguments about the nature of the universe (and other theories which will come along) it should be remembered that all of them are speculative. They are based, however, on other well-tested theories like the Theory of Relativity; the absolute constant of the velocity of light; Hubble's constant (recession velocity = distance); the Doppler Effect, with its red-shift significance; gravitation; non-Euclidian geometry of curved space; and many, many, more. Those insights have suggested observations, and observations have provided fresh insights. All go into the stockpot from which the cosmological chefs prepare their recipes. If someone were radically to challenge Einstein's theories (many have tried and many keep on trying); if, somehow, it were found that the light constant was not universally true; that Hubble's constant had to be discarded; and that there is a quite different explanation of the red shift, much of the vast amount of observations and information on which the contending theories are based would have to be drastically revised. Even the paradigm of the Expanding Universe would be as doubt-ridden as Anaxagoras's churn. It is presumptuous even to suggest it, but 'That's the way the cosmic cookie crumbles.' Meanwhile it is still the framework within which to judge newly discovered facts, like quasars, or anything else which is around the corner of the curved universe.

A do-it-yourself cosmology

So far we have discussed only the general theories of the universe. Let us now have a look at the actual, observable patterns on the cosmic wallpaper.

We shall accept (as an unsolicited gift) the Stuff of the Cosmos. We have got matter mobilized and we choose our own D-Day and we give the word 'Go.'

Our Stuff of the Cosmos is '*ylem*' (cf. Aristotle), meaning 'the first substance from which the elements were supposed to be formed'. And *ylem* came out hot and strong – 15,000 million degrees Absolute. This figure can be arrived at by accepting the present mean temperature, not of a room, but of the universe, as 50 degrees Absolute; by applying a formula based on the relation between Hubble's constant and the mean density of the universe, and tracking back to H-Hour of our D-Day. At that temperature, the particles were rushing around with the velocity of those contained in the modern atom-smashing machines. Matter must have been in a state of 'plasma', in which no composite nuclei could exist – a 'soup' of elementary particles, but a 'soup' so tenuous as to be a gas formed of protons, neutrons, and electrons.

As soon as *ylem* began to cool off, neutrons began to associate themselves with protons and electrons and the first atoms began to assemble. The whole time for the aggregation of nuclei (for nuclear-physical reasons we shall not go into) was one hour. The density of *ylem* was, however, extremely high, so that the polygamous neutrons must have married and remarried. As the temperature dropped to a mere 10,000,000,000 degrees Absolute, the light atoms of hydrogen were formed and, in the next phase, the heavy atoms.

Nobel prizewinner Enrico Fermi and Anthony Turkevich explained what happened in the first thirty-five minutes after *ylem*'s jailbreak. In the first five minutes the temperature of the universe was still too high to permit the creation of complex nuclei, so that all that was happening was the philandering of neutrons with protons and electrons. With a further lowering of temperature, neutrons and protons married to form deuterium nuclei (double-hydrogen) which almost immediately doubled up to form helium. In thirty minutes, almost half the original *ylem* had become hydrogen and slightly less than half had become helium. In reality (because it does not exist in anything but rare quantities) most of the deuterium must have been swallowed up in the building of heavier nuclei. Anyhow, in the first hours of the universe, the roll call of the atomic species was present and correct.

Then *ylem* went on recuperative vacation for 30,000,000 years.

The hot gas continued to expand and, in the process, cooled off to only a few thousand degrees Absolute. The various elements with high melting points congealed into a fine dust, floating in a mixture of gas and helium. This kind of mixture reveals itself to us today in interstellar dust, which causes the absorption lines in the spectrographs, and the drag on the wavelengths of distant stars which accounts for the red shift. This dust accumulates as giant clouds or strange shapes called either 'luminous' or 'dark' nebulae, to the degree they are illuminated by near-by stars.

How did this cosmic dust begin to form into galaxies, stars, and planets? There was a lot of matter, but there was also a lot of radiation, which took the form of photons (already discussed), and (in anticipation of 1905 and Einstein's $E = mc^2$) energy had mass and could push around atoms. But gradually energy which had been pushing around matter began to diminish and matter began to assert its rights of free association. But matter is subject to the laws of gravity, which acted on the particles scattered by the photons almost uniformly through space. As Sir James Jeans showed, a gas subjected to gravitational forces and spread through unlimited space is unstable and will break up into separate giant gas clouds. This condensation of primordial gas could form clouds the smallest possible mass of which would be several times that of our Sun. This critical minimal mass is interesting because it is calculated from purely nuclear data and suggests an analogy, which is more than coincidental, between the macrocosmos of the celestial system and the microcosmos of the nuclear particles. In looking (as we are in this discussion) for a unified pattern throughout the universe of matter, this is at least a freehand drawing, an enlargement of which is the Cosmos and the reduction of which is the atom.

Those proto-galaxies were moving outward and they were spinning. They had some difficulty in disengaging from each other because of their mutual gravitational attraction. They were like rockets which have the necessary escape velocity but are being braked by the gravity of the Earth until they finally get clear.

Depending on the degree of rotation obtained by the proto-galaxies in the separation process, their gaseous bodies assumed different shapes. If the rotation was low, they could form spheres;

faster, they would become ellipsoid, the elongation being a factor of the rotational speed; very fast, the bodies flattened into a lens-shape, like a yo-yo, with material streaming out of it like the jet from a catherine wheel. These latter, the most common, are the spiral galaxies. The outstretched 'arms' turn out to have a much less significant role in the structure of the galaxies than was once assumed. They are formed of dusty gas streamers, which, caught by the general rotation, have got twisted into spirals like a Christmas festoon. The main bulk of the galactic disk consists of a vast number (billions) of stars-in-their-own-right rotating in regular circuits around the hub.

The origin of those stars (including our own Sun and the Milky Way) has naturally been the subject of much observation and speculation. The motion of the gas masses in the proto-galaxies must have been quite rugged. The gases near the perimeter had a tendency to rotate with less angular velocity, a longer period of rotation, than the inner masses. As in our own directly observed planetary system, the rotation periods increase with the distance from the Sun at the centre, but whereas the planets are self-contained, and separated, the situation in a gaseous system would be more like a fast-flowing river going round a continuous bend. On such a swerve the smooth flow breaks up into a series of small-scale, irregular motions which churn along with the main flow.

This turbulence is of enormous practical importance in hydrodynamics and in aerodynamics, from the effects on the banks of rivers to the drag on aircraft wings. The characteristics are well-recognized in wind-tunnel experiments where the introduction of smoke gives a clear picture of the eddies swirling past an obstacle, e.g. a wing section of an aircraft. Turbulence has been put on a strictly mathematical basis by some of the best thinkers in science. Briefly, the motion of eddies involves the release of kinetic energy which is transferred from larger swirls to smaller ones. The smaller the eddy, the smaller its velocity and the shorter its life tends to be.

Applying these observations of turbulence, as we see it on Earth, to the gaseous proto-galaxies in our changing universe, the German physicist and cosmologist Carl von Weizsäcker has reconstructed what probably happened. In the kinetic theory of gases the internal friction (viscosity) of gases is given by a product of gas density, thermal velocity of the gas molecules, and their free path between successive collisions. Velocity differences between parts of rotating gaseous galaxies could have been at least as large as ten kilometres per second, whereas the thermal velocity of gas molecules at the cooling temperatures prevailing at the time was probably less than one kilometre per second. The mean free path of the molecules in the highly diluted gas was calculated to have been as much as 100,000,000,000 kilometres. This looks enormous but it is negligibly small compared with the dimensions of a galaxy. The combination of these factors represents conditions in which the motion of the gases probably became turbulent, producing a breakup into eddies of all sizes. This, in a gaseous state, would produce compression, or condensation. Without Newtonian gravity, this condensation would disintegrate, but in the gravitational fields existing in the proto-galaxies, instead of expanding and again streaming off, the local condensations should have continued to contract under their own weight and form dense gas spheres. As a consequence, the temperature of the gas spheres would have risen steadily and have begun to radiate from their surface heat waves and then waves of visible light. As the contraction increased, the central 'furnace' of the spheres became so hot that the conditions of thermal fusion began, and the nuclear reactions produced a powerful source of nuclear energy.

These are the stars which we see around us today. A great deal has been observed, and deduced, as to what goes on inside them. Since no spaceman in an asbestos, radiation-resistant suit will ever be able to explore the interior even of our own star, the Sun, we shall, forever, have to rely on earthly analogues. We shall, of course, be able to extend our senses and our instrumental methods to check whether the answers at which we arrive are consistent.

'Steady state' appears to have been discounted by the observed facts, although 'continuous creation' is not necessarily so. An 'expanding universe' is not inconsistent with a 'contracting universe' if we accept the idea of a 'pulsating universe,' i.e. in which the process of the 'big bang' may be repeated. This, obligingly, provides us with elements synthesized in secondary stars, which gets over some of the difficulties that cosmologists have in accounting for the more complex elements which had to go into the making of, for instance, our own solar system.

5 Stars and Planets

Solar nuclear reactor

Much of what we now know about the Sun derives from the work, in the twentieth century, on the nucleus of the atom. Basic to those studies is thermonuclear fusion, what first we reproduced on Earth in the cataclysmic form of the H-bomb.

As has been indicated, nuclei get fused, or welded together, when temperatures are sufficiently high. The heat of the interior of the Sun is 15,000,000°C. Even higher orders of temperature are produced at the instant of the explosion of a fission bomb, with its artificial fissile material, either U^{235} or U^{233} or plutonium. This is used as the percussion cap or detonator of the H-bomb, which is surrounded by a wadding of material rich in deuterium (double-hydrogen) or tritium (triple-hydrogen) which can combine in a quick build-up to the combination of four hydrogen atoms to make helium. The excess energy is violently released. In this explosive process man has taken short cuts to get results in split seconds. The Sun does its 'cooking' of atoms in a more leisurely manner (it has billions of years). A convincing cycle of changes was first proposed by Hans Bethe. In this cycle the carbon nucleus acts as a catalyst, taking part in the reaction at the beginning of the cycle but reappearing with the formation of the helium nucleus.

The cycle is as follows: the carbon (C^{12}) captures a proton (the hydrogen nucleus) and forms an isotope of nitrogen (N^{13}). The nitrogen isotope decays into a carbon isotope (C^{13}) by emitting a positron and a neutrino. Carbon-13 captures a second proton, forming ordinary nitrogen. The nitrogen next captures a third proton to form an isotope of oxygen (O^{15}). This immediately decays by emitting a positron and a neutrino and becomes another nitrogen isotope (N^{15}). In the final stage the nitrogen isotope captures a fourth proton, resulting in the formation of a helium nucleus and an ordinary carbon nucleus.

But a second process, closer to that of the H-bomb, is going on in the Sun at the same time. Two protons combine to form a deuteron emitting a positron (positive electron) and a neutrino (a particle with zero rest mass and half spin). The deuteron captures a proton to form an isotope of helium (He^3). Two helium isotopes now combine to form ordinary helium, discarding two excess protons. Three plus three (He^3 plus He^3) make six; subtract two and leave four (He^4), which is stable helium.

Astrophysicists, through spectroscopic analysis, know how much hydrogen is available in the Sun (or any other observed star), and they know the rate at which the hydrogen is being consumed by conversion into helium.

Calculations show that the supply of hydrogen in the Sun will last for about 5,000 million years.

(One recalls the story of the lecturer who quoted this figure. A member of his audience rose in great agitation and asked him to repeat the figure. 'Five billion years,' said the lecturer. 'Oh, thank goodness!' sighed the interrupter, 'You had me worried. I thought we had only five million years.')

Red Giants and White Dwarfs

To get further insight into the internal economy of the stars, we should look at two other types which are extremely interesting – the Red Giants and the White Dwarfs. As we have seen, the nuclear reactions which convert hydrogen into helium and so keep the star supplied with energy depend on temperature; the higher the temperature, the fiercer the reaction. Since the core of a star is much hotter than the rest, hydrogen there will be devoured, while in the outer 'atmosphere' the hydrogen will only be nibbled, or unaffected. So the central regions will become helium-rich, while the layers near the perimeter will remain pure hydrogen. The radius of the star will increase by the heating of the outer gases and the star will become a Red Giant. (If this were to happen to our Sun it would grow steadily more incandescent and this would make the oceans boil on Earth, which would presently be swallowed up, and the Sun would go on expanding until it would devour the other planets.) The transition to gigantism occurs first among the most massive and brightest stars (the Sun is medium). In the expansion stage, the heat at the core can sustain the outer surface at a much lower temperature. This, optically, gives the star its red colour.

The White Dwarfs appear to be exactly the opposite of the Red Giants. They are examples of very bright stars which have burned up more and more hydrogen until they have eaten up their atmospheric hydrogen as well. Then they are not able to maintain their heat balance. They begin to shrink. This is accompanied by a cooling process which causes the constituent particles to become more and more compacted until they cannot be compressed further. From dimensions many times that of our Sun it will become a ball no bigger than the Earth, and extremely dense. A piece of White Dwarf material no bigger than a matchbox would weigh over two tons. It is no longer generating heat but the concentrated matter is white hot and emits a lot of heat and luminosity per unit area. It has a steely blue appearance. When it finally cools off, it becomes a Black Dwarf.

This is a plausible life history of the stars, but as has also been indicated, they can have even more extravagant adventures, like the supernovae.

Providing the planets

'Give me matter', wrote Immanuel Kant, in *Allgemeine Naturgeschichte und Theorie des Himmels* (*General Natural History and Theory of the Heavens*), 'and I will construct a world out of it.' And he proceeded to do so from a recipe no longer entirely acceptable by modern astrophysicists.

Kant's philosophical assumptions of the origins of the planets were given a scientific plausibility by Laplace, at the end of the eighteenth century, with his nebular hypothesis. Laplace argued that at some remote epoch all the material which forms the Sun and the solar system was a vast nebula of rarefied gases in slow rotation. The nebula cooled slowly and, as it did so, the gas contracted because of the internal gravitational forces. The rotation quickened as the contraction proceeded. Eventually, the rotation became so fast that wisps of the gases became detached from the periphery and formed 'smoke rings' outside the nebula. This happened over and over again, as the rotation increased, until the central portion of the nebula condensed as the Sun and the material of the 'smoke rings' collected into separate planets. This very conveniently accounted for the distribution of the planets in the same plane. Two fatal objections were raised by Clerk Maxwell a century later. He argued that the 'smoke rings' could never have coalesced into large single planets. They would have formed into smaller bodies, like Saturn's rings. The second objection (which continued to be fatal for later theories as well) was concerned with the momentum of the planetary system: a large mass has more momentum (i.e. a larger quantity of motion than a smaller one moving at the same speed). (Momentum is, in fact, calculated by multiplying mass by velocity.) Since the planets are moving in orbit around the Sun and not in a straight line, theirs is a rotational momentum.

One point about which there is no disagreement is the measurement of the masses and velocities of all the planets of the solar system. The Sun weighs a thousand times more than the Earth. Indeed, 99·9 per cent of the mass of the solar system is in the Sun. If the masses of the planets are multiplied by their velocity, it should be that nearly all the rotational momentum should be carried by the Sun. Not so, exactly the opposite is the case: 98 per cent of the momentum is carried by the planets. This is because the planets are moving very fast in their orbits compared with the rotation of the Sun. This separation of mass and velocity must have occurred when the peripheral rings separated from the main nebula. If 98 per cent of the momentum of the system was concentrated in those rings, then the velocities must have been enormous: so great, in fact, that even small droplets, let alone planets, could not have formed.

The Kant–Laplace hypothesis was undermined by this authoritative argument. Two American astronomers, Chamberlin and Moulton, proposed in the twentieth century another type of origin, to escape from the embarrassments of the nebular theory. Their hypothesis was that our Sun was originally an ordinary star without planets. Then, about 20,000,000,000 years ago, another star passed close to it in their common journey through space. The gravitational attraction led to a temporary flirtation and then the other star passed on its way. In this close encounter great tidal waves of matter would be torn from the Sun. Some of it would be sucked back; some of it would follow the other star, but some of it would remain in the gravitational field of the Sun, circling around it. The blobs would congeal and merge as they cooled, like candy floss being spun by rotation. This tidal theory had continuing fascination. Jeans and Jeffreys varied the tidal theory by suggesting that the passing star pulled out from the Sun a great filament of gas which (using again the crude analogy of candy floss) broke up and congealed immediately, without any progressive accretion process. This in turn was invalidated by the American astronomer, H. N. Russell, who showed (1935) that in any such encounter, the passing star would have had to pass so close that the planets would have to be many thousand times closer to the parent Sun than they are today.

Otto Schmidt and the astronomers at his institute in Moscow tended to compromise with Kant and Laplace by suggesting that the planets originated from a diffuse cloud of material surrounding the Sun. In their theory, the cloud consisted of both dust and gas and the planets were formed by the gradual accretion of those cold particles. As this huge primordial cloud rotated around the sun the dust particles gradually concentrated into a flattened disk. In innumerable collisions between the primeval dust particles, the particles lost some of their velocity, the energy being discarded as heat, until appreciable aggregates began to form. The embryos grew into substantial bodies, with diameters of hundreds of miles. A tremendous turmoil existed with collisions occurring in which many of the giant boulders got smashed. But all the time the accretion process was 'snowballing' into bigger and bigger units, with those which remained intact growing into the major planets. The theory suggests that the Earth was not one of those primary embryos but was formed from fragments which acquired more dust particles. This, and the rounding of the mass into a sphere, by spinning, required a few thousand million years.

Space exploration has shown that the bombardment of the earth with cosmic dust is still going on at the rate of millions of tons a year, but the dust does not reach the surface because our atmosphere forms a dust sheet. That is why we have to go outside, by satellite, to find out what is happening. We still get meteorites puncturing the dust sheet but, by and large, we are not so dusty.

Meteorites which reach the Earth (becoming incandescent through friction with the atmosphere which burns most of them up) are our only direct experience of ultra-terrestrial material. They are believed to come from the region of the asteroids. These smaller bodies are in the gap between Mars and Jupiter. One theory ascribes their origin to the break-up of a planet which existed between the two planets. Alternatively they could be survivors of those primordial embryos which failed to mature. Only a few of them are as much as a hundred miles in diameter, but the number increases as the size diminishes and there is no obvious distinction between small asteroids (boulders as big as a mountain) and the size of the giant meteorite which crashed in Siberia in 1908. A great deal of work has been done on the structure of meteorites, and the results appear to tally with the suggestion that they passed through successive stages of accretion –fragmentation–accretion.

This accretion idea overcomes most of the difficulties of the nebular theory of Kant–Laplace. But even here there are differing ideas: either the dust cloud was captured by the Sun, at a time when the Sun was already self-sufficient as an ordinary star or the Sun and the nebular cloud were born at the same time.

Otto Schmidt and the Russian school sustained the first idea that the Sun got tangled in the cloud when it passed through a dense region of interstellar dust. Hoyle at one time favoured a modified version. He suggested that the Sun was once a member of a twin system, a binary star. These double stars revolved around each other in close gravitational affinity. One of the Siamese twins (according to Hoyle, at this stage of his thinking) exploded and, although the 'mushroom' carried away most of the debris, enough remained to form the gas-dust cloud from which the planets accreted.

Later Hoyle considered that the Sun was born as one member of a large cluster of stars and in the beginning the Sun and the cloud were coexistent. Hoyle suggested that one of the clusters of stars associated with the preplanetary sun was in fact a supernova which exploded, and elements which accreted into planets were the fallout of this super H-bomb explosion.

The buts in rebuttals

To show how even the most stern rebuttals can, in turn, be qualified by further evidence, let us consider what happened to Clerk Maxwell, whose rejection of Kant–Laplace had promoted the collision (or rather, near-miss) and tidal theories. In 1944, Carl F. von Weizsäcker showed that Maxwell's objection could not hold because of what had been discovered in the interval about the chemical constitution of the matter forming the universe.

In Maxwell's time it was believed that the Sun and the stars and the interstellar material had the same constitution as Earth, being generally composed of silicon, iron, oxygen, and a few other elements forming different abundant materials. Astrophysics turned somersault when it was found (spectroscopically) that this was not true and that the terrestrial elements form less than 1 per cent of the composition of the Cosmos. The remaining 99 per cent consists of a mixture of hydrogen and helium. Weizsäcker argued that the original solar nebula must have had a much larger mass than that contained in the planets as they exist today, the excess of hydrogen and helium having been lost after the planetary bodies were formed. With the total mass of the original material thus increased a hundredfold, Maxwell's argument fell: it would have been possible for the planets to have formed through condensation.

These revised ideas were further revised by G. P. Kuiper. He estimated that the original nebula around the Sun possessed a mass about 10 per cent of the solar mass and that its breakup, caused by gravitational forces, could have led to individual condensations, each with the same constitution as the original interstellar material and only 1 per cent (or less) of what are now the most abundant terrestrial elements, then in the form of fine dust. The dust particles could have gradually sedimented towards the centre to form the cores. The interesting feature of the Kuiper theory is that the formation of the nebulous planets took place in darkness. The Sun was not yet effulgent. As the Sun contracted almost to its present size, its surface temperature went up several thousand degrees, and it began to pour out radiation. The photon pressure of ultra-violet waves scoured off the hydrogen-helium envelopes of the planets and they would look very much as comets do today, with luminous tails trailing behind their bodies. Mercury, Venus, Earth, and Mars, being closest to the 'hose' of ultra-violet, had their atmospheres practically stripped off and they were left as solid bodies with only a membrane of primordial atmosphere. (The now much denser atmosphere of the Earth came much later as oxygen and nitrogen leaked from the rocky crust.) The outer planets Jupiter and Saturn did not lose much of their hydrogen-helium envelope, and their bulk today is really gaseous with relatively small cores in the centre.

The secondary satellites around the planets can similarly be accounted for: just blobs caught in the gravitational forces. A special argument can be made for the Moon. Compared with the satellites of other planets it is comparatively large. Although it is only an eightieth the size of the Earth, it is possible (and Moon investigations will prove or disprove this) that they were (like binary stars) simultaneous creations and that they were much closer together. The complete absence of atmosphere on the Moon could be explained by the 'scouring' by ultra-violet rays, but it is also true that the Moon's gravity is not sufficient to keep a hydrogen-helium envelope around it.

Life on other planets

As will be seen from all this, theories about the origins of the solar system are many. They come and go and sometimes they come back. As they say at the fairground, 'You pays your money and you takes your choice!' When theorists and their protagonists say, 'It happened this way...' or use 'did' or 'must', they have no right to be so categorical. They are being tentative, however thoroughly they do their homework, and they base their ideas not on proved but temporarily approved evidence.

One thing which appears to be generally agreed is that the planets did not originate from the molten state, like bullets from boiling lead. Whichever of the current theories you pick for the 'Top of the Pops', the analogy would be rather with powdered metallurgy, in which the granules of metal are compressed together, and the planets would be accounted for by accretion, under pressure, of cold, solid particles. The high temperatures that exist in the Earth's interior, for instance, arose later because of the tremendous pressures which developed and because of the heat produced by the radioactive processes (fission, not fusion) of the heavier elements.

> In the final stages of accretion [wrote Sir Bernard Lovell] the outer layers remained cool, and if any complex molecules or organisms already existed in the dust cloud they retained their identity [cf. Anaxagoras].[1]

We will consider later this 'prebiotic dust' and the conditions under which life, in whatever form, can exist. One may dare to say categorically that the Earth is not unique. Indeed, modern cosmogony, no matter from which ideas or common body of evidence the various theories of the origins of our planetary system are derived, would accept the situation in which most of the stars in the Milky Way have planetary systems like our own and that the same would be true in all the other galaxies within and beyond the reach of our telescopes.

That would give us billions of possible earths and, with what some would call 'mathematical certainty', millions of worlds at precisely the same stage of development as our own. In this time-coincidence, they too would be sending out space probes to investigate their neighbours in their own planetary systems. Few scientists were entirely derisive when Russian radio-astronomers announced in 1965 that, among the consistent signals they were getting from stellar sources, there was a rare frequency on which there was a repetitive signal. The first suggestion that this was 'someone' in another 'world' in another universe trying to get

1. *The Individual and the Universe*, Oxford University Press, 1959.

through was hurriedly qualified into a 'maybe'. It is, nevertheless, a reasonable 'maybe'. The signal *could* be consistent with a radio call sign. How we would 'get the message' is another matter. With a humility that is difficult to accept, we should consider that there may be beings more advanced than we are. In other galaxies the conditions of the evolutionary process towards creative intelligence could have been anticipated or could have happened faster. We say 'Ninety per cent of all the scientists who have ever lived are still alive.' That means that 10 per cent are sparsely spread along the corridors of time back to the first masters of fire. If we can have a concentration of scientific talent, with its chain reaction of discoveries, in our own lifetime, why should not some other planet have had its breakthrough at, say, the time of Aristotle? The reason we have not registered their existence may be that they are millions of light-years away by radio. Or maybe we are not clever enough to read the message!

6 The Envelope

The Earth we live on

Before we consider the Earth sciences, i.e. those which deal with the nature of the planet on which we exist, let us go back to the Sun which, we might agree, brought Earth into existence. One thing about which we cannot dispute is that it is the source of our continued life.

Let us briefly identify it. It is the nearest of the stars, a hot, self-luminous globe (as distinct from the planets which shine by reflected light). It is, by comparison with other stars, only medium in size. It contains 1,000 times as much material as Jupiter, the biggest planet, and over 300,000 times as much as the Earth. Its gravitational attraction regulates the performance of the planets and its rays supply the energy which maintains nearly every form of activity on the Earth. The temperature at its surface, 440,000 miles from its centre, is 6,000°C., which is twice as hot as that achieved in an electric furnace. But the temperature in the interior is 15,000,000°C. This is, as we have seen, the heat generated by nuclear action. This produces radiation in the form of gamma rays, which are powerful, penetrating X-rays. Fortunately for us, the radiation does not reach us in this form; otherwise life could not exist. The gamma radiation is battered around from atom to atom as it struggles to the exterior of the Sun, like someone being buffeted around trying to get out of a crowd. A gamma ray which starts off now will not reach the Earth for another 20,000 years (remember that the Sun's light rays reach us in eight minutes). During this wandering around in the interior of the Sun, the gamma radiation is successively absorbed and re-emitted until it has become the tolerable sunlight which we enjoy. This flow of energy, this 20,000 years wandering in the solar wilderness, finally ends in a spurt through the Sun's atmosphere. This lasts only a few seconds, but before it leaves the Sun forever each atom in the Sun's surface gives it a message. These separate signatures of separate atoms are what we spell out as the lines in the spectrum of the Sun.

The Sun's blanket thus moderates the heat. If it were to unmuffle ten times as much heat, the brightness would increase a thousandfold. The Earth's oceans would evaporate and the rocks would melt and bubble like porridge. The Earth is also protected by a blanket of atmosphere which controls the intensity and character of the rays bombarding it. We have become familiar with this kind of protection because we are being told all the time about the way in which the space satellites are measuring the ultra-violet radiation, the cosmic rays (the atomic particles from space), the meteors, the dust, and so on. From the full and damaging impact of all these, our atmospheric blanket protects us.

The sea above us

We live at the bottom of a vast and deep ocean of air. Our atmosphere consists of about four-fifths nitrogen, one-fifth oxygen, and small amounts of other gases. It also contains water vapour, which may vary in amounts to a little more than 4 per cent of the total volume. The total weight of the atmosphere gives a pressure of fifteen pounds per square inch at the Earth's surface.

Until the mid twentieth century about as much (or as little) was known about the behaviour of this 'ocean' as was known about the currents of the seven seas. In spite of the elaborate system of weather stations and weather ships, meteorology was an inexact science and weather forecasting (as the television weathermen kept apologetically reminding us when their predictions went wrong) was rather hit-or-miss. That was because, without a great deal more information, plotting the weather was freehand drawing, like Leeuwenhoek's sketches of his 'little animals' as he saw them with his primitive microscope.

The meteorologists were dealing with forces generated by the Sun. Although only two-thousand-millionths of the Sun's total energy reaches the Earth, this is sufficient to provoke the disturbances which give us the manifestations of our weather. Weather remained only precariously predictable because there was at best only a sum of local observations dominated by local surface features of the globe, mountains, and so on, which, though very significant day by day in the localities involved, have as little effect on the general atmospheric system as rocks at the bottom of an ocean have on the oceanic currents.

The two main factors affecting the motion of air masses over the surface of the Earth are: (1) the uneven heating of the polar and equatorial regions, which creates the driving forces for atmospheric circulation, and (2) the rotation of the Earth around its axis, which causes the deflection of air streams moving from the poles to the Equator and from the Equator to the poles. Moreover, oceanography can never be dissociated from meteorology, because the oceans and the atmosphere together form a heat engine. The waters of the oceans are evaporated by the Sun, and redeposited as precipitation which finds its way eventually back into the seas. The cold and hot currents of the oceans interact with the atmosphere, chilling or tempering it, and influencing the pattern of precipitation.

One of the objects of the International Geophysical Year (1957–9) was to repair the shortcomings of superficial meteorology. One of the successful projects, in cooperation with all nations and with scientists of all nationalities, was to create a weather fence around the globe. This coordinated system of observations meant that 'the weather' at and near the surface was systematically 'sieved'. But the 'highs' and 'lows' of the daily forecasts are only items. There are many other variables in the weather equation: notably hurricanes, those cataclysms which move so quickly and strike so disastrously. To be able to forecast hurricanes and save human lives, if not property, it is necessary to witness their birth (anywhere within a million square miles of ocean) and observe their movements, and to study and derive the physics of these movements: the fantastic amount of energy represented by their swirls and their swerves. To get that kind of information it is necessary to be outside looking in. With a system of weather-eye satellites it is possible to get pictures of all the hurricane-generating systems every twenty-four hours, in much the same way as, at a much lower level, helicopters can watch traffic movements. Whereas earthbound weather observatories, with radiosonde balloons and even rockets, can get only fractionalized information, satellites can give accurate pictures and relay back detailed information about the upper air.

Before meteorological science could claim to be exact, great advances had to be made in the physical understanding of the general circulation of the atmosphere and the weather systems to which it gives rise. The weather satellite provides means of discerning instantaneously and continuously whole sectors of the upper atmosphere where the 'highs' and 'lows' are predetermined. Its instruments can tell the forecaster exactly how much energy is available in the upper atmosphere for the manufacture of wind, cloud, and rain and the day-to-day variation of the amounts which will affect the weather on which so much of our comfort and material needs (crops, etc.) depend.

Above the troposphere, that part of the atmosphere characterized by turbulent mixing, cloud formation, and, generally, a decrease in temperature with altitude, with an upper boundary varying from 25,000 to 55,000 feet, there is the stratosphere. This is the region of the jet streams, powerful atmospheric currents, in a steady temperature. Above this is the mesosphere, the hot ozone layer, where temperatures increase with height up to 160,000 feet and then decrease from there to 250,000 feet. Above this is the ionosphere where the temperature again increases with elevation. In such conditions, and with bombardment by shortwave radiation from the Sun, molecules dissociate into their atomic components and, footloose and fancy free, become charged ions. The ionosphere contains two principal regions of free electrons. The lower (E, or Kennelly-Heaviside Layer) occurs at heights of 55 to 75 miles and the second (F, or Appleton Layer) is found from 100 to 200 miles above the Earth. These layers are influenced by the Sun and their characteristics change with the time of the day, the season of the year, and the sunspots.

The sunspots are phenomena of the Sun's surface. They are clearly visible to the ordinary observer through densely smoked glasses as an umbra (dark centre) surrounded by a penumbra (a smoky fringe). They appear to move across the face of the disk, but this motion is due to the rotation of the Sun. Individual sunspots seem to be capricious, appearing for a few hours or a few months, but when the 'population' of sunspots is studied, there is a regularity and consistency which expresses itself as a minimum and a maximum. The cycle takes eleven years, from maximum to maximum. At the periods of intense activity when, apparently, the spots flare out like volcanoes belching nuclear particles and radiations, there is interference with the ionized layers around the Earth. This can lead to the blackout of radio communications.

The nature of sunspots and the mechanism of the sudden onset and development of solar flares are among the tantalizing problems which scientists are always studying. So is the structure of the chromosphere; so is the unstable transition region which surrounds the bright disk of the Sun (the photosphere), and so is the substance of the corona, the Sun's tenuous outer atmosphere, which may fill the whole of interplanetary space.

Much can be done in studying the Sun from Earth-located observatories. The U.S. National Science Foundation financed a giant solar telescope on Kitt Peak, Arizona (7,000 feet above sea level), with a focal length of 300 feet, which gives thirty-four-inch diameter images. But knowledge about the Sun is restricted by the opaqueness of the 'window', the atmosphere, through which we look at it. It is necessary to get out of the sea of air for a clear look. This is what satellites and space probes can do. And that was what the planners of the International Geophysical Year proposed. Their ambitions were modest in comparison with the 'space spectaculars' which followed.

Getting the information

The International Geophysical Year (I.G.Y.) was an exercise in scientific cooperation unparalleled in history. More than seventy nations joined forces and the committees of those nations were responsible for the operation of 2,500 major stations as well as many thousands more temporary stations. More than 30,000 scientists, engineers, and observers gathered vital information – vital in the sense that what they were examining determines the existence of life on Earth. It was a civilian enterprise managed by the International Council of Scientific Unions (I.C.S.U.) and sponsored by the United Nations Educational, Scientific, and Cultural Organization (Unesco). An eighteen-month period (1 July 1957, to 31 December 1958) originally was designated as 'International Geophysical Year', directed towards a systematic study of planet Earth and its cosmic environment. This period was later extended to 31 December 1959. The measurements undertaken fell into three major categories: (1) physics of the upper atmosphere, which included measurements of solar activity and particles from the Sun (then at sunspot maximum), the stars, and the interplanetary medium; (2) the heat and water regimen, which included meteorology, oceanography, and glaciology; (3) the Earth's structure and interior, which included seismic, gravimetric, and longitude and latitude refinements.

Outside looking in

In the preliminary planning (1954) I.G.Y. recommended that satellites should be included. Both the U.S.S.R. and the U.S. responded. It was generally assumed that the United States with its acknowledged technological supremacy and its engineering capacity would be first to put a satellite in orbit.

On 4 October 1957, however, the Soviet *Sputnik 1* was launched. It weighed 184 pounds and carried instruments to determine and relay back to Earth data on temperatures and pressures.

In that instant, space research became politics. The U.S.S.R. was accused of having 'jumped the gun', but while the details of its launching dates had not been publicized, the Soviet scientists had, in fact, kept I.G.Y. headquarters in Brussels advised of the 'phasing'. It could be deduced from that information that if all went according to plan the Soviet countdown would be some time around October 1957. Sputnik's inclination to the Equator was 64° 3′, which meant that, in the rotation of the Earth, every inhabited place was traversed by its orbit; that could be exploited for propaganda advantage, such as telling the astrologically minded Indians that any child born under the zodiacal sign of

Sputnik would be privileged. The U.S. was planning an orbit at inclination 33° 5′ to the Equator (round the cummerbund of the globe) in a west–east trajectory that, on the slingshot principle, could 'borrow' roughly a thousand miles per minute – the speed of the Earth's rotation. The Soviets did not thus economize and that, in conjunction with the 184-pound capsule (compared with the 30·8-pound U.S. capsule), showed that the Soviets had employed a very powerful launching rocket.

Many were surprised by the Soviet achievement, although it was no longer thought that, in technological terms, they were *moujiks* who tore the entrails out of imitation Ford tractors. They had shown manifold engineering skills and, like the Americans, had acquired a quota of the German rocket experts who had produced the V-2s of the Second World War. Moreover, in 1903, Tsiolkovsky, a Russian, had been the first responsible scientist in the world to propose large-scale space rockets, liquid-fuelled. Nor was the orbit a surprise to anyone who knew the traditional scientific aptitudes of the Russians, distinguished long before the Revolution in mathematics, astronomy, ballistics, and chess. Nor, for a country which had produced Pavlov, was it surprising that *Sputnik 2* (launched 3 November, 1957) with a capsule weight of 1,120 pounds, should have included a dog to give information about the physiological and neurological effects of orbital flight, as well as instruments to measure cosmic rays, solar ultra-violet and gamma radiation, temperatures and pressures.

Nevertheless, this initiative by the Soviet Union, followed on 31 January 1958, by the U.S. *Explorer 1*, turned scientific cooperation within the I.G.Y. into fierce competition in the cold war. The weight and the unassisted orbit were taken as proof that the U.S.S.R. had Intercontinental Ballistic Missiles (I.C.B.M.s) and to spare. This aroused fears of 'the missile gap' and resulted in the stepping up of military rocket programmes. The innocence of the I.G.Y. satellite programme disappeared in the threat of 'space mines', surveillance 'space stations', and 'bases on the Moon', of a new military space strategy.

Together with military activities, however, there was an incomparable increase in civilian space research. Within a few years hundreds of satellites were in sustained orbits (apart from those scheduled for limited objects with limited life-spans), and men (and a woman) had adventured in space. The teething troubles had been overcome and the space engineers could design and place satellites more or less at will – like the weather satellites, and communications satellites on a commercial basis to universalize television. Until satellites were available, television transmissions had been restricted to horizon range, but now the very short waves could be reflected from man-made 'mirrors in the sky', so located in orbit and in equilibrium with the speed of the Earth that they remained 'fixed'.

Aiming at infinity

The Moon became a target. In terms of space engineering, launching an Earth-orbit satellite or a 'shot' to the planets or the Sun is a matter of no great difference. As one can see, as a televiewer watching the apparently leisurely takeoff of the launching pad 'booster' and the buildup of speed by the successive stages, the final escape velocity depends on the height at which the final thrust is made. At an altitude of 3,950 miles (one Earth radius) the gravitational pull (consistent with Newton's Laws) is only 25 per cent of what it is at the Earth's surface. At 23,700 miles it is down to 2 per cent. At 250,000 miles it is practically nonexistent. The Moon is, on an average, only 239,000 miles from Earth. So a shot to the Moon is a shot to infinity. What has to be contrived is that, in shooting to infinity, the Moon, Venus, or Mars gets in the way. This is rather more difficult than shooting flying geese, but it is the same kind of anticipation of position.

Not surprisingly, the first attempts at planetary marksmanship were either failures or near-misses. Some of them failed to escape from the Earth's gravitational attraction. The Soviet rocket fired on 2 January 1959 took two days to come abreast of the Moon and missed it by 5,000 miles. It became the first man-made satellite of the Sun. An American attempt, two months later, missed by about 35,000 miles and also sped on into orbit round the Sun. On 12 September 1959, *Luna 2*, the Soviet probe, making the journey in 33 hours, 32 minutes, crashed on the Moon but, in its 'last will and testament', revealed that the Moon was enveloped by a layer of low-energy ionized gas. On 4 October 1959, *Luna 3* was launched and later transmitted television pictures of the hindside of the Moon, which had never been revealed to Earthlings before. The Soviet Union on 12 February 1961, launched a probe scheduled to reach the planet Venus in about 100 days. Its radio transmitters failed about 4,700,000 miles from Earth after measurements had been relayed back about interplanetary magnetic fields and solar radiation. The first news from Venus was transmitted by U.S. *Mariner 2* launched on 27 August 1962. The planet, which is about the size of Earth, is veiled in thick clouds.

The unveiling of Venus by *Mariner 2* was a remarkable example of how scientists can extend their senses. To rendezvous with the planet the probe had travelled a distance of 182,000,000 miles: a cruise almost half way round the Sun. It reached its nearest point to the Sun – 65 million miles – on 27 December, four months after launching. It was then 2,700,000 miles from Venus, and 44,213,000 miles from Earth. It was under control from the Earth and had been redirected when it was obvious at one stage that it would arrive at the rendezvous ahead of Venus, missing it by 233,000 miles. A rocket motor was fired for twenty-seven seconds, sufficient to reduce the velocity and bring it within 21,000 miles of the target, well within the intention. Moreover, its built-in electronic brain had failed to operate the radiometers by which the essential information was to be obtained. The command was restored by direct radio signals from Earth. The instruments began to work and continued their probing of the concealed planet for forty-two minutes, after which the planet moved beyond the range of the scanners.

The magnetometer had recorded the magnetic fields which it had encountered in interplanetary space but, in its fly past, failed to record any Venusian magnetic field. The planet either had no magnetic field or a very weak one. This finding had important bearings on the origin of magnetic fields, the rotation of Venus, and the temperature at its surface, with the question mark about the existence of conditions which might preclude the possibility of life. The magnetic investigations suggested that the high temperatures which ground observers had noted from the measurements of radio waves from the planet existed only in the upper layers of the atmosphere and are produced by an intense bombardment of cosmic rays which are diverted from Earth by its magnetic field. The second important message from *Mariner 2*, however, sent back details of directly measured temperatures showing that the temperature at or near the surface of Venus is between 300° and 400°F. This would rule out any possibility of life even remotely resembling that on Earth.

Another dramatic 'first' was the United States tryst with Mars on 14 July 1965. Across a distance of 137,876,000 miles, twenty-one photographs of a surface sweep of Mars were sent back to Earth. They were taken at heights varying from 9,500 to 10,500 miles above the planet. The camera scannings were stored on magnetic tape – like taking notes and transcribing them later. On playback, the data for each picture took eight and a half hours for transmission to the Earth. The signals were received on high-speed printers which expressed them as a series of numbers. On the continuous strip of paper two columns of digits combined to form numerals from 0 to 64, representing gradations of shade. A computer then transferred the information to magnetic tape which reconstructed a film version of what *Mariner 4*'s camera had recorded. The results were like a tapestry, but showed topographical features of Mars – a pockmarked planet – such as no earth-bound telescope could ever resolve. In addition, stored information about the nature of Mars was relayed back. Among other things it showed that Mars's mass is about one-tenth that of Earth. The indications were that if life existed, it would have to survive in a pressure environment equivalent to Earth altitudes of at least 93,000 feet – over three times higher than Mount Everest.

The magnetic fence

Although the space programme acquired a different meaning and momentum when it moved out of the academic atmosphere of the International Geophysical Year and into the bear pit of cold war politics and strategy, the scientific results were prodigious. (Whether they were worth $10,000,000,000 a year is a social evaluation which belongs, not with this discussion of Nature, but with consideration of the direction and purpose of science.) A great deal of information about the know-how of space engineering was, of course, withheld, but, in terms of I.G.Y. (and Cospar, the International Committee on Space Research, which derived from I.G.Y. as a continuing body) there was an abundance of data to share. At that level cooperation continued. The United States and the Soviet Union made such 'unclassified' information available to each other and to the world. The United States provided facilities – acted as Wells Fargo for instruments – for other countries and cooperated with the nations of Europe which set up the European Space Research Organization (Esro).

Apart from masses of data over which scientists will work for decades, certain 'first-sight' discoveries were made. For example, it was known that the Earth bulged at the Equator and was flattened ('like an orange', one was told at school) at the poles. Before satellites were launched scientists had made estimates of the bulge from theoretical calculations, based on the centrifugal force of the Earth's rotation. Observation of the satellites in Earth orbits showed that those calculations had been remarkably accurate but that the Earth was shaped not like an orange but like a pear. From the minor variations in the satellite orbits, due to differential gravitational Earth pull north and south of the Equator, it was shown that the South Pole is about 50 feet nearer to the centre of the Earth than was thought, while the North Pole is 50 feet farther away. The Northern Hemisphere is about 25 feet nearer the centre than had been thought, and the Southern about 25 feet farther away. Considering the size of the globe, variations by about the length of a dining hall show the minute difference between the deduced and the observed.

Another new discovery was the Van Allen Belts, named after the American scientist who planned the satellite experiments which confirmed their existence. At the beginning of the twentieth century, Carl Störmer, a Norwegian physicist, had deduced that the charged particles which are sent out in all directions from the Sun were trapped in the Earth's magnetic field in a large ring around the planet. In this ring the movement of the charged particles would produce a massive electric current. This current would be surrounded by its own magnetic lines of force. These lines of force reaching down to the surface of the Earth would be part of the magnetism which influences compass needles. Man-made satellites confirmed that Störmer was right. The first clues were found in 1953 when some once-for-all rockets were fired into the air near Newfoundland. The instruments of the rockets showed that there was an unusually large amount of radiation thirty miles above the ground. This finding was confirmed by *Sputnik 3*; whenever it came into the northern auroral region its instruments showed a sharp increase in the measurement of X-rays which came from the shell of the capsule when it was bombarded from outside by fast-moving electrons.

The most important evidence came from the U.S. *Explorer* satellites and *Pioneer* space probes. *Explorers 1* and *3* had particle counters which showed very high counts of radiation at 600 miles. The *Pioneer* space probes passed right through the heart of the belt and added to the picture by showing that there were two radiation zones, one within the other. The inner ring is doughnut-shaped and extends to about 2,000 miles from the surface of the Earth; the outer one extends from about 8,000 to 17,000 miles from the surface of the Earth.

Then in 1962 came surprising news from *Pioneer 5* that the radiations extended to, and a disturbed magnetic field existed at, distances from 40,000 miles to 60,000 miles and that the interplanetary boundary of the Earth's magnetic belt was twice as far from Earth as the farthest previous assumptions. One of the direct consequences of this discovery was a newfound hazard to interplanetary travellers because in passing through the magnetic zone they would be exposed to radiations of from 10 to 100 roentgens *an hour*, whereas the radiation which a human being can tolerate is 5 roentgens *a year*. This meant extra precautions.

But the discovery of the radiation belts introduced a scientific as well as a human hazard. The original observations of the Van Allen Belts were made in January 1957. On the basis of them, Project Argus was proposed. It involved exploding nuclear bombs in the Van Allen region to release charged particles. The proposal evoked protests from distinguished scientists all over the world, because this involved 'contamination of space' with consequences which no one could foresee and with the risk of interfering with further observations. Political protests followed. The

Prime Minister of Great Britain was asked to intervene with the U.S. Government. He replied that his experts were not as alarmed as the protesters and, anyway, what was all this concern about a Belt which a year before no one had known had existed? Three bombs were exploded in August and September 1958. The charged particles were observed to bounce back and forth along the lines of force, and, after arching out above the Equator, returned to the Earth's atmosphere to produce man-made auroras.

Effects of space litter

The scientific world had been sensitive about 'litter' from the beginning of space research. An emergency committee, Cetex, 'contamination by extraterrestrial exploration', was set up in 1957. This was later absorbed as a responsibility of Cospar. It was foreseen that Man's intervention might interfere with pristine conditions. (Astronomers had been hounded outwards and upwards, moving their observatories because they found their observations being more and more obscured by urban smoke and the ground glow of mercury and sodium street lamps.) But it was not just the question of inconvenience for the scientists. It was recognized that more serious consequences could arise. Claude Bernard's nineteenth-century injunction that 'Science proceeds from the known to the unknown' was counsel to go warily, consolidating each step by attested knowledge. There is always a great temptation when someone has a bright idea with novel means available to try it out quickly and with the chance of establishing priority. The motives are never described that way. The intentions are always of the highest. In Houston, in September 1962, President Kennedy said:

> We choose to go to the Moon in this decade and do other things, not because they are easy but because they are hard; because that goal will serve to organize and measure the best of our energies and skills; because that challenge is one that we're willing to accept; one we are unwilling to postpone, and one we intend to win...

Which recalls one of the earliest preoccupations of Cetex. It was pointed out that the Moon, without atmosphere to serve as a dust sheet and without winds, had been accumulating cosmic elements (a sort of celestial grand piano) and that nothing was known about that dust. It could, however, be in a 'prebiotic state', with aggregations of elements which might lend themselves to reactions which had never been triggered off. This was a time of intense preoccupation with D.N.A. (deoxyribonucleic acid), which will be discussed later, and with the possibility that this could produce a chain of chemical events if it were introduced from Earth. Sceptical derision was accorded suggestions that the Moon might be a sort of laboratory bench and that biological processes might be started off. No one suggested that life, as we know it, would spring, like Minerva, from a capsule, but there were those who foresaw blights being introduced from Earth, say, a living ooze, or pus. Men-in-haste pooh-poohed such misgivings, but the warnings had some restraining influence. Spacecraft intended for direct hits were sterilized, lest the sneeze of the last mechanic to turn the last bolt might introduce a germ culture of a D.N.A. sequence. In the excitement over the pictures of Mars from *Mariner 4* an item of small-type insignificance was the statement that the probe had been routed to pass not less than 5,000 miles from Mars, to avoid any risk of a collision because the utmost biological precautions had not been taken; the last stage was not completely sterile.

As Sir Harrie Massey, chairman of the British Council for Scientific Policy, wrote in *The World in 1984*,

> ... the question as to whether large-scale modifications of the Earth's environment may not have been brought about [by 1984] which will have introduced serious extraterrestrial sources of interference, not only for radio-astronomy but for other scientific and general pursuits. [By that time] the problem in this direction will have reached such proportions that a great deal of international discussion will be concerned with the setting up of a suitable system of controls to minimize dangers arising in this way.[1]

This was a recognition by responsible scientists, for whom Massey (a 'space-man' himself) was speaking, that 'extraterrestrial contamination' was not just a question of 'fouling up'

1. Ed. Nigel Calder, Penguin Books, 1965, vol. 1, p. 34.

scientific observations, present and future, but a threat to the conditions of life on Planet Earth.

Apart from satisfying man's insatiable curiosity, which no one would want to discourage (except the Faustians, who would say that such inquisitiveness invites eternal damnation), what 'practical' purposes can interplanetary travel serve? Academician E. K. Fedorov of the U.S.S.R. was optimistic at the United Nations' Conference on the Application of Science and Technology for the Benefit of the Less Developed Areas (Geneva 1963). The world population is increasing at an alarming rate. The United Nations demographers estimate at least six billion Earthlings by the year A.D. 2000 (and the emphasis is on 'at least' – if death control is not offset by birth control). That did not disturb the academician. He foresaw the other planets of the solar system being invoked (as the New World was invoked to redress the population problems of the Old World in the nineteenth century) and providing living room and raw materials for Earthlings. The idea of immigrant rockets, celestial *Mayflowers* to the Moon, Mars, Venus, or points beyond, is fascinating but unreal. The total emigration from the old countries to the new, throughout a century and a half, was 50,000,000, substantially less than the *annual* population increment nowadays. At a cost of $15,000 per pound (*avoirdupois*) to send anything into space, the freight charges for, say, a round trip to get raw materials to Venus would be rather excessive!

There is what space research economists call 'fallout'. In getting what someone called 'a $20-billion handful of Moon dust', technological problems had to be cracked (with expense no object) and the results could be converted to terrestrial industry. One of the most important was the use of solar energy to provide continuous power for the space vehicles' transmitters. With the 'write-off' of research and development costs, the cheapening of solar devices could be of immense benefit to many power-starved, sun-baked, developing countries. The crash programmes to get man hurriedly into space provided biologists, in the space-medicine research laboratories, with means to study human beings and to gain knowledge of importance for the earthbound as well as the space adventurers.

But the direct and immediate values were in terms of weather and communications, which gave the world's inhabitants better opportunities to control their environment and, for good or ill, to familiarize themselves with each other, like neighbours gossiping (or quarrelling) over the garden fence.

The stimulation of interest in the possibilities of life on the solar planets (which had been imagined by science fiction writers for so long but was not subject to direct examination) encouraged research into the conditions in which life (not necessarily identical with ours) could arise. The debate on such origins will be developed later.

7 Planet Number Three

Man-made satellites circling the Earth sixteen times a day have reminded us how relatively small Planet Number Three is (Number Three in distance from the Sun – Mercury, Venus, Earth, Mars, Jupiter, Saturn, Uranus, Neptune, and Pluto), and have given us a sense of proportion about our place in the universe. In the celestial domain, the stars (of which our Sun is one) are thought to be more numerous than the grains of sand on all the beaches of the world, and billions of them have planets.

But what the eyes, human or electronic, in orbit see is the family estate of the human race, the individuals of which are increasing by the billions. The optic mosaic, like the television camera mosaic, registers a picture of a slightly distorted sphere, like a classroom globe in sculptured relief but not a very emphatic relief because from, say, 130 miles up Mount Everest's 29,000-foot height is a pimple, the great mountain ranges are just wrinkles, and the Grand Canyon is a scarcely visible crease. They register the superficial features of a globe which is seven-tenths oceanic water and three-tenths land. Of that three-tenths (the continents and islands) about two-fifths are covered by hot and cold deserts and mountains, one-third by forests, one-fifth by permanent pastures, and one-tenth by cropped lands.

So the rangelands and cultivated lands on which the inhabitants of the Earth depend for their sustenance are a minor part of the 'family estate'. The satellite eye cannot see what is in the crust of the Earth, where men have to find the fossil fuels and mineral resources on which our technological civilization has become compulsively dependent. Nor can it see the depths and bottom of the oceans, which submerge seven-tenths of the planet and about which we know less than we know about the visible surface of the Moon.

The Earth cannot be considered distinct from its envelope: the magnetosphere, which extends to between 40,000 and 60,000 miles from the centre of the Earth; the ionosphere, the reflector ceiling; the stratosphere; and the troposphere. We can also differentiate the atmosphere, the air component; the biosphere, the environment of all living things; the lithosphere, the solid surface; and the hydrosphere, represented by water.

The natural atmosphere (as distinct from the man-made atmosphere, e.g. smog created by imperfectly combusted fossil fuels and chimney effluents) consists of nitrogen, oxygen, argon, carbon dioxide, neon, helium, krypton, xenon, hydrogen, methane, and nitrous oxide, and this gaseous envelope is influenced by the biosphere since living things transpire and, in the process of living and dying, convert and liberate chemicals from the soil. Soil may be basically a product of the rocks, eroded by the rains and winds, themselves the violent forces of the Sun-driven atmosphere, but the soil is also 'living'. A lump of soil no bigger than a football contains more organisms than there are people on Earth, and humus, the decay product of living materials, is an essential component of the topsoil on which all growth depends. In the topsoil organisms break down the living materials and release the chemicals which plants need as nutrients, and make the 'crumbs' of soil which hold moisture and also help the roots to 'breathe'. From this topsoil grow the plants which provide the food for animals and humans and at the same time provide the vegetation texture which protects the soil from erosion.

All these are subject to energy radiations generated by the Sun, which by remote control dictates the environmental conditions of our planet. Only to a small extent is the lithosphere so influenced: by the surface forces which create erosion, carve and grind the rocks, and by the deposition of the carboniferous layers of primeval living materials, which have stored the Sun's energy as hydrocarbons.

For the rest, the crust of the Earth and its core are self-contained. If we accept the tenet that Planet Number Three, like the others, was formed out of nebular gases and dust condensing, fragmenting, and again accreting, and reject the theories that it was extracted from the Sun, it had a separate creation, although reared as one of the solar family.

Our planet has an outer crust about twenty miles thick, an egg-shell compared with the dimensions of its envelope. The crust is extremely thin below the oceans. Beneath this is the 'mantle', and beyond that the central core. It is like a golf ball: the dimpled casing (the crust), the different layers of elastic windings (the mantle), and the fluid sac in the centre (the core). The Earth has a radius of about 4,000 miles. One of the earlier discoveries of seismology, the study of earthquakes, was the discovery of the central core and its dimensions.

There are two types of wave which travel from the site of an earthquake downwards into the Earth. They are the primary or P. waves in which the particles of the Earth's material move longitudinally as the shock waves pass by, and the secondary or S. waves in which the particles move sideways. In solids, P. waves move 50 per cent faster than S. waves. P. waves take 20 minutes and 12 seconds to go right through the Earth to the other side, diametrically. S. waves do not transmit through fluids to any significant extent. Therefore, those parts of the interior found to transmit both P. waves and S. waves are solid. Failure to detect S. waves is evidence that the part of the Earth involved is in a fluid state.

Seismology located the central core 1,800 miles below the surface. This gave it a radius of 2,200 miles. In 1936, this was further defined by the seismological discovery of an inner core of 800-mile radius. Thus, apart from the crust there are three zones in the interior of the Earth.

Seismology mainly derives its information from natural earthquakes, although man-made explosions are used in geophysical prospecting of the crust of the Earth and can give localized insight into the nature of rock structures and determine the presence of oil or water. Megaton hydrogen bombs (equivalent to millions of tons of T.N.T.) were exploded at Eniwetok in the Pacific in the spring of 1954 and these, and others, produced earth tremors that could be registered at great distances, showing that the shocks had penetrated the deep interior of the planet. This was table rapping compared to the ten or more major earthquakes which shake the Earth every year. The earthquake which occurred in Assam, India, on 15 August 1950, had the energy of about 100,000 atom bombs.

Larger quakes can be recorded by seismographs in all parts of the world. If the receiver is far enough away from the site of the quake, the disturbances reaching it will have travelled through the planet's deep interior. From this kind of data scientists can learn a great deal about the structure of the Earth.

The lower boundary of the crust is called 'the Mohorovičić Discontinuity', after its Yugoslav discoverer. Below this is the mantle extending to the boundary of the outer core at a depth of 1,800 miles. The outer core is fluid (i.e. in a molten state) and the inner core is dense, most likely iron, solidified under the extreme pressures existing at the centre of the Earth. Seismology can determine the velocity of shock waves, which vary with depth. There is a progressive stiffening of the material according to increasing pressure. In the mantle the pressure rises rather regularly at a rate of about 470 atmospheres per kilometre. (One atmosphere of pressure is 14·7 pounds per square inch.) The pressure at the boundary of the core is 1,370,000 atmospheres and at the centre it is 3,700,000 atmospheres.

As we have seen, the aggregation of the planet by 'powder metallurgy' from the nebular dust and fragments has, as a theory, replaced earlier speculations about its origin as a ball of molten lava erupted from the Sun. The materials from which the planet was formed must have been relatively cold. It could have been 'toasted' by the Sun's rays, but that would not have generated sufficient heat to account for the interior structure. The original Earth, throughout billions of years, was (on present assumptions) non-molten. It could have come near to melting at a later stage as a result of the heat of the radioactivity generated by the fissile atoms inherent in its own structure. At some time the Earth must have been soft enough to allow the separation of the various layers which the seismologists can distinguish.

Radioactive substances generating heat tend to associate with the lighter elements which are mainly present in the outer layers, notably the crust. This would account for the solid centre, the homogeneity of the lower outer core, and the non-homogeneity of the upper parts of the outer core. This would be like cooking a baked potato in a radioactive foil or like a 'Baked Alaska' – the ice cream remains solid in the centre, while surrounded by a filler, with crisp meringue on top, cooked by the radiant heat of the oven.

Chemical sampling of rocks can give direct insight into the elemental nature of the Earth's crust. The findings can be compared with the chemical analyses of the meteorites which puncture the 'dust sheet' of our atmosphere and which, as has been pointed out earlier, may be the fragments of a planet, or embryonic primordial planets which failed to mature. Another comparison can be made with the chemical elements of the Sun, revealed by spectroscopy. And this spectroscopic analysis can be extended to the other stars. The results from all the examined sources – Earth, meteors, Sun, and stars – are reasonably consistent and support the contention that the gross chemical composition of universal matter is uniform. 'Cosmic abundances' of elements have been estimated, using silicon as the reference element, because rocks consist largely of compounds of silicon (from Latin *silex*, meaning flint). What carbon is to organic chemistry, silicon is to inorganic chemistry – it enables complex molecules to come into existence.

Suess and Urey (1956) worked out the relative abundance of elemental atoms in the universe. For every atom of silicon there are 40,000·0 atoms of hydrogen; 3·5 of carbon; 0·91 of magnesium; 0·6 of iron; 0·37 of sulphur; 0·15 of argon; 0·09 of aluminium; 0·05 of calcium; 0·04 of sodium; 0·03 of nickel; and 0·01 of phosphorus. Only six of the remaining elements have abundances which lie between a thousandth and a ten-thousandth of the availability of silicon. The others exist only in minutest traces.

The Earth, when forming, lost most of the volatile gases – unlike Planet Jupiter. Some of them had, however, combined to form non-volatile compounds, such as water (hydrogen and oxygen), or the silicates (silicon with oxygen, magnesium, and iron), or the nitrates, compounds of nitrogen and oxygen with metals. The gases certainly do not exist on Earth in anything like the proportions given in the 'cosmic abundance' table.

Geologists are able to distinguish between rocks which belong to the surface layer and magmatic rocks extruded from the interior in molten state. Magma must be similar to the material of the mantle. The oldest extruded rocks are largely composed of olivine. The commonest form of olivine contains about one part of iron to nine parts of magnesium in association with silicon and oxygen. But the proportions can vary and it is most likely that the amounts of iron increase the farther down in the mantle. The seismic evidence agrees convincingly (i.e. with laboratory tests) with the assumption that the Earth's mantle consists mainly of minerals of the olivine type. Since examination of the meteorites also shows this type of mineral to be dominant, it could give support to the theory that meteorites are fragments of a planet which had consolidated itself by heat and pressure and had acquired an interior similar to Earth's, in preference to the suggestion that meteorites are nebular embryos, or minuscular planets. These would not have the necessary heat and pressure to provide the autoclave for production of olivine.

Earth's outer core is molten iron. There may be traces of nickel and other 'impurities', but nearly pure iron is consistent with physical and chemical data. The explanation of the solid iron core is plausibly simple: the heavy iron particles 'sedimented' towards the centre. That, of course, would also be true of elements heavier than iron, but the proportion of those elements in the Earth is quite inadequate to fill the solid inner core, which has a radius of 800 miles, as was shown, in 1936, by a Danish woman seismologist, Miss I. Lehmann. Further evidence, including the analyses of the shock waves from the Eniwetok hydrogen explosions in 1954, supported her calculations. There were two estimates of the density at the centre of the Earth: one giving it a value of eighteen grammes per cubic centimetre; the other only sixteen grammes per cubic centimetre.

Earthquakes usually show patterns of oscillations on the order of a second. The Kamchatka earthquake of 1952 registered oscillations of 57 minutes and 100 minutes. Chaim Pekeris, at the Weizmann Institute in Israel, put a computer to work. It showed that these oscillations were possible only if the density at the centre were eighteen grammes per cubic centimetre.

The giant dynamo

William Gilbert was Queen Elizabeth I's royal physician and as such officiated at her deathbed. But he has been conferred with immortality for a greater reason: his discovery that Planet Earth is a big magnet. Magnetism had been known long before his time; e.g., by the Greeks (the name derives from Magnetes, the inhabitants of Magnesia in Thessaly); by the Chinese; by the Moslem traders; and by the Crusaders, navigating to the Holy Land. But it was Gilbert who defined the dip and declination of the compass needle and attributed the magnetic forces to the Earth itself, thus explaining the pointing of the needle north and south towards the poles (a term which De Maricourt, a French Crusader, had applied to the places from which magnetism appeared to originate).

Gilbert, with the limited information at his disposal, originated a paradigm which was supported by all relevant observations after his time, including the findings of space research.

The question was whether the magnetism, of which the Earth was unquestionably the magnet, originated inside or outside the planet. Johann Carl Friedrich Gauss (1777–1855), usually acknowledged as the founder of modern mathematics, showed that there were two sources, inside and outside.

Subsequent studies showed that the internal magnetic field could not originate in either the crust or the solid mantle of the Earth but only in the fluid metallic core. The movement of that fluid core generates electric currents. Eddies in this fluid produce the irregularities which express themselves as observed variations in the Earth's magnetic field at the surface. But the fluid motion not only modifies the electric currents, it generates them.

Just as electric current is produced in generating stations by the rotation of metallic wires past each other, the streams of molten metal passing each other in the fluid core 'scrape off' each other's electrons and liberate them as flowing currents. The motion of the core can be explained by thermal convection. This is like water in a kettle which is heated from below; the bottom water, as it heats, bubbles to the top. In the case of the Earth this is a globular rotating kettle.

No one seriously questions the fact that during the last 500,000,000 years (at least) the magnetic field of the Earth has been consistent with its present one. This is borne out by the study of rock magnetism, a fairly recent science that originated in an attempt to establish a unity between gravitation and magnetism, which, incidentally, was suggested by the behaviour of a binary (dumb-bell) star. To do certain instrumental checks, the scientists went down into deep mines and became extremely interested in a magnetic manifestation in rock structures. They found natural 'compass needles' embedded in certain formations.

These take the form of crystals of iron oxide which retain the 'remnant magnetization' of the magnetic field to which they were originally exposed. When sedimentary rocks were deposited at various times, iron oxide bore this magnetic imprint. It is as though a compass needle floating in a bowl of water were frozen in the direction of the pole to which it was pointing at a given moment. A study of these 'pointers' in rocks all over the world has produced some extraordinary evidence. In successive layers of sedimentary rocks magnetic crystals are differently oriented. There are three possibilities: first, that the axis of the Earth has shifted – even somersaulted – so that the North Pole could have been at the South; second, on the dynamo theory, that the electric currents of the self-excited generator, Planet Earth, have repeatedly reversed their direction; third, that the crust of the Earth has shifted.

By plotting the directions indicated by the remnant magnetization of the iron oxide crystals it is possible to determine where the rocks were in relation to the magnetic poles (which do not coincide with the geographic poles). Judging by those plots, the North Magnetic Pole must have wandered from somewhere in the region where Japan is at present to its present position somewhere in the region of Prince of Wales Island north of the Canadian mainland. By the same token, from evidence of rocks in the Southern Hemisphere, the South Magnetic Pole must have been a vagrant too. Magnetic rock crystals have been found in the Northern Hemisphere which point to the South Magnetic Pole. They have 'changed sides', across the Equator.

The third possibility – the creeping of the Earth's crust – would be quite consistent with this evidence. It would mean that the skin, whether intact or rifted, must have moved so that it was the land masses, containing the rocks, which were wandering and not the magnetic poles.

Continents adrift

Rock magnetism would, therefore, strongly support the Wegener paradigm of 'Continental Drift', which had nearly foundered in controversy. It was restored to respectability, if not complete acceptance, by the testimony of the rocks.

The theory was propounded in 1912 by Alfred Wegener (born in 1880, died on the Greenland Ice Cap in 1930). He suggested that the land mass was coherent – one vast island, in the one primeval ocean – and that it broke up. The fragments became rafts of sial, the uppermost layer of the Earth's crust, floating on the heavier sima, the basaltic layer of greater density.

Wegener's theory had a superficial plausibility to anyone who looked at the map. Indeed, it was its simplicity which made sceptical scientists suspect it. If one frets out the outlines of the continents, bearing in mind that we are dealing with a sphere and not a flat Mercator projection, one finds that most of them, properly manipulated, fit neatly together like pieces of a jigsaw puzzle. For example, the bulge of Brazil fits plausibly into the Gulf of Guinea in the continent of Africa. Restoring the continent of North America to Europe is not quite so simple. Wegener, however, insisted with reasonable justification that the break-up had been like that of an ice field, in which the floes would not only separate but would tend to slew. Given that, one can manipulate the Spanish peninsula and Newfoundland so that they can be made to fit (remembering, too, that one should take account not only of the supramarine coastline but of the submarine continental shelf). In the Arctic, the picture can be easily restored by repacking the land floes of the Canadian archipelago, Greenland, Spitsbergen, Novaya Zemlya, etc. Moreover, if one extends the Caledonian mountain formation of the Scottish Grampians and of Wales, one finds a persuasive conformation with the Appalachians in the United States. In the eastern and southern hemispheres the fitting together becomes even more intriguing. If Madagascar is tucked back closely to the continent of Africa, and one assumes that Australia slewed when it broke away, the Australian Bight can be made consistent with the contours of South Africa, with interesting relevancies between the gold reefs of Australia and those of South Africa. When the fragments of Indonesia are swept together and the seam bursts of the Red Sea and the Persian Gulf are restitched, there still remains however a gap where the Indian Ocean is – the missing continent of 'Gondwanaland', the Oriental equivalent of 'Lost Atlantis'. If, however, one regards the Scotia arc (the Falkland Islands and the South Shetland Islands) at the tip of South America as a hinge, one can close South America into Africa and swing the Antarctic continent into the gap in the Indian Ocean. This lends a fascination to future studies of Antarctica, because under the ice cap of that frozen continent would be the geological wealth of rock structures which once bound East Africa to south-east Asia. It would explain the coal seams of Antarctica – relics of once-Equatorial forests.

On this hypothesis, the greater land mass of Eurasia has remained more or less anchored, with some substantial internal stresses which account for the squeezing up of relatively new mountain ranges like the Himalayas, and with the Americas breaking away and shifting westward in response to the centrifugal force of a globe which is spinning west to east. This provides a rational explanation of the western coastal ranges of North and South America – the Rockies, the Sierras, and the Andes. Those great upthrust ranges could be the 'bow waves' caused by the sial pushing against the resistance of the sima. And it accounts for many of the characteristics of this region, like the volcanic faults and the phase slipping of the geological structures lending itself to grinding shifts which reverberate as earthquakes.

This westward shift of the continents adrift is also supported by the evidence from the submarine volcanic arc of the mid Atlantic, stretching from the polar seas, down the North and South Atlantic, and round the Cape of Good Hope. One can take as markers the volcanic activities in Iceland, the Azores, and in Tristan da Cunha, but these are only surface peaks of a submerged system. And the extraordinary thing about this system is that it consists of twin, parallel ranges with a continuous gap between. This rift and a crack in the sima are detectably widening. Thus Wegener's concept of sial and sima has posthumously received substantial geological and seismological support. The sial, or raft of relatively lightweight rock which supports the continental superstructure, is approximately five to ten miles thick. This is resting, or floating on, a heavier, darker, crystalline material which wraps the world around and extends down to an average depth of twenty miles. This layer (largely basalt) is the bedrock of the oceans, but is depressed by the weight of the continents.

The inside pushing out

The molten magma from the interior intrudes through the sima and the sial and through vents and fissures in the formations of the crust. If the vents are vertical the effect is like a boil or carbuncle which erupts as rounded or conical volcanoes. If the fissures are lateral the comparison would be with a laceration with the lava oozing and forming scars (like the mid-Atlantic ranges). Sometimes the exudation gets trapped in the surface layers, forming pockets of molten materials which, cooling, form the rough granites, which may eventually be exposed by erosion. Or the gases may be trapped in pockets and may escape through 'fumaroles', holes or cracks in the surrounding rocks. The escape may be as hot gases, or it may take the form of hot water or steam (geysers), from the heating of groundwater encountered en route. The huffing and puffing of those geysers (like Old Faithful in Yellowstone Park, Wyoming) are due to the intermittent action of the gases. In the liquid magma the bubbles are lighter than the enclosing fluid and rise through it, like the bubbles in porridge and, like porridge bubbles, smaller bubbles merge to form bigger bubbles. Considerable pressure can build up in the bubbles when they encounter resistance, and explosions varying in degree from a puff to a violent eruption may precede the lava flow of a volcano. Sometimes they are no more than the plop-plop of the mud pools.

Volcanoes are a general feature of the Earth's surface. Some have been extinct so long that their original activity has been forgotten and the regions in which they developed are no longer regarded as volcanic. But volcanoes have always been a principal artificer in shaping the Earth's topography. The 'workshop benches' of the world were the so-called 'shields'. The Pre-Cambrian formation (long before the earliest vestiges of life appeared) characterized the interiors of continents. (From the radiation measurements, already described, some of those primary rocks must be 4,500 million years old.) From, and around, those 'benches' new features and new mountains grew. The sedimentary rocks were deposited by erosion and compression and the Ice Age glaciers spokeshaved them into new forms. The benches cracked and parts shifted. Earthquakes and movements in the Earth's crust jolted them around. Volcanoes erupted and formed mountains which in time the elements whittled (by erosion) and the movements of the glaciers altered into shapes which do not remind us of cones or domes. Igneous rocks, such as the granites, were extruded from within. But around the benches long after they became stable grew new mountains of more recent volcanic origin. And it is characteristic of new mountains that alongside them on the seaward side are immense deep troughs, or trenches (Mindanao Deep, off the Philippines, is 6·7 miles deep; it would drown Mount Everest). Those new mountains, the overspill from inside the Earth, are subject to severe erosion, and the material slides like an avalanche into the trenches and the sediment piles up on the slope (the geosyncline) like the spoil on a slag heap. That sediment is compressed into rocks, and in the course of hundreds of millions of years the consolidated debris gets pushed above sea level (as the 'bow wave' of the Wegener movement?) to form the folds of the mountain ranges. Contemporary evidence confirms that even the rocks of the 'shields' are more ancient at the centre than towards the margins. And the Rockies, the Sierras, and the Andes, along the line of volcanic fracture, are so young that their shapes have not been worn down by the sandpapering of time.

Thus volcanoes and earthquakes are not only the catastrophes of our headlines but the plastic surgeons of the world's lineaments. Yet the scientists who call themselves 'volcanists' or 'seismologists' are very low in the scientific hierarchy, humoured as enthusiasts, like the butterfly hunters. They may get more recognition by calling themselves 'geophysicists', implying that what is really their basic interest is a sideline. There has been little recognition of the natural heirs of Pliny the Elder, who died of sulphur suffocation investigating the eruption of Vesuvius (the path to Hell), which overwhelmed Herculaneum and Pompeii, and of Empedocles, who, according to the poet Matthew Arnold, threw himself into the crater of Etna to justify his scientific principles. (There are other legends of his death).

Far more attention should be paid to those phenomena – to forces indigenous to the Earth and not derived, like other energy effects, from the Sun. In this at least – Inner Space as contrasted to Outer Space – Earthlings ought to be parochial. And for reasons other than righteous inquisitiveness, because it can have practical value.

Geothermal heat

If one looks at the volcanic arcs of the world, their configurations are significant to countries which are power-starved. One arc stretches down the whole the of Western Hemisphere, including Latin-American countries which are looking for development. Another stretches from Kamchatka, through Japan, the Philippines, westward through the Moluccas, Lesser Sundas, Java, and Sumatra. Another begins in New Zealand, through the New Hebrides, the Solomons, the Bismarck archipelago, and the islands north of New Guinea. Another, more reticent, volcanic system extends along the Great Rift Valley of East Africa, northward through the deep cleavage of the Dead Sea. A transverse system extends from the active volcanoes in and around Italy and through the Middle East (Ararat, where the Ark is said to have landed, is one of the ancient volcanoes) to Iran. And the Atlantic twin ranges have been mentioned.

The first serious attention to those arcs as a practical source of geothermal energy was drawn by the United Nations Conference on New Sources of Energy (1961) held in Rome. One of those 'new sources' was geothermal energy, deliberately extracted from the volcanic formations. Local exploitation had long been maintained in, for example, Iceland, with its geysers. The heat from those had been used to generate electric power, and to supply domestic heat and for an efficient greenhouse horticulture economy. Reports were also presented of remarkable achievements in power generation at Larderello, near Pisa, in Italy, where hot gases generated enough electricity (and valuable chemicals were also recovered from the gases) to supply the equivalent of all the power needed to run the Italian state railways. New Zealand had harnessed the underground sources of geothermal steam at Wairakei in North Island for electricity generation. The geysers of California had become an automated process of electricity generation. These were all in association with 'open showings', familiar vents or eruptive steam or mud pools. What the geothermal engineers had to say, however, was of worldwide importance: that artificial plumbing into geothermal formations could provide even greater heat than the natural vents; that is to say, contiguous, locked-in heat (as gas or steam) could be released with up to twenty times the temperature of the fissures, or pools or geysers, due to pressurized heat. This, it was suggested, could be of great economic value to many countries geothermally endowed, but lacking fossil fuels or hydroelectric potential. It had the advantage that the goethermal source, once identified, could be tapped where economically suitable – for siting of industries or towns. The power units could be as large or small as need be. It would be like drilling a well and servicing local industry, without the high capital and transmission costs of, say, a massive hydro-electric dam. The United Nations offered technical assistance to any member states interested in this 'new' (surely the oldest?) source of terrestrial energy, but found it difficult to recruit experts in this largely neglected subject to carry out the surveys.

Into the interior of the Earth

Jules Verne (1828–1905), impresario of so much of the public entertainment which became billion-dollar business and international politics in the second half of the twentieth century, made his *Twenty-Thousand Leagues Under the Sea* and his *Journey to the Centre of the Earth* even more exciting than his *From the Earth to the Moon*. The submarine *Nautilus*, which made the first long-range journey under atomic power and went under the polar ice cap, was named for his fictional ship of 1873. But the fulfilment of his journey to the interior lagged a long way behind Man into space and Man to the Moon. Even with atomic submarines regularly circumnavigating the globe, and with human beings returning in the popular sport of skin diving to the oceans from which life emerged, probing Inner Space did not command the lavish appropriations which went into Outer Space.

When the Mohole depth probe was first proposed, the deepest that living man had gone was in a bathyscaphe descent of seven miles into the Marianas Trench in the Pacific by Jacques Piccard and Don Walsh (23 January 1960).

In the mood of scientific adventure created by the International Geophysical Year, accompanied by the excitement of the first space satellites, the Inner Space men saw their opportunity. 'Mohole' was proposed as a project to bore through the boundary between the Earth's crust and the mantle. As has been pointed out, the sima, which supports the continental sial raft, is the bedrock of the oceans. The samples recovered in the course of drilling would give a new insight into the structure of the Earth, at least in its upper layers, as well as a great deal of information about the origins of life.

Preliminary trials were authorized and were carried out in 1961. A specially designed ship was used to conduct drilling in 3,000 feet of water off La Jolla, California (the site of the Scripps Institution of Oceanography), to a depth of 1,035 feet into the ocean floor. A second experiment was tried in two miles of water, and the drill penetrated 601 feet into the layer. This second experiment was a variation of the offshore oil-drilling techniques extensively employed: drilling from anchored platforms (instead of a ship, as in the first Mohole venture) into oil formations in the continental shelf.

What was to be looked for was evidence which would enhance knowledge of the nature of the planet. The objective of drilling to the mantle was to make it possible to obtain samples of the upper crust and of the chemical (or petrological) composition of the mantle. Mohole drills were to bring up cores which would be analysed chemically and physically as to their mineral character, radioactive content, density, and conductivity, both thermal and electrical. Given precise information about the density of the containing rock, the estimates of density all the way to the centre of the Earth could be more accurately determined. When the radioactivity of the sample could be measured, it would be possible to say whether the Earth was cooling and explain the high flow of heat through the ocean floor. But, as the then director of the project, Willard Bascom, said in presenting the scientific prospectus, 'The most important objective is to look for the unexpected – unpredicted discoveries which are valuable because they upset accepted theories.'

The determination of the nature of the sediments through which the drill would pass to reach the Moho could be obtained by taking a continuous cylindrical sample – a vertical cross section. This could help to answer a number of bewildering questions. The rate of deposition on the ocean floor has perplexed scientists. The estimated thickness of the first layer of sedimented crust (on top of the sima) is 1,600 feet on average. It is calculated that the average deposition of the fine sediment which could be spread uniformly over the ocean bed is one centimetre every 1,000 years. Thus it would take only a few tens of millions of years to deposit a layer comparable to the first layer. How then does it come about that, after all the billions of years of ocean existence, the first layer seems to represent less than 100 million years? Either the estimates are far out or something infinitely strange occurred at some time in the ocean depths which completely altered the history of the original ocean basin.

8 Seven-Tenths of the Globe

During the planning of the Normandy landings in the Second World War, Winston Churchill issued an imperative instruction to the Chiefs of Staff

> Piers for use on beaches: They must float up and down with the tide. The anchor problem must be mastered.... Let me have the best solution worked out. Don't argue the matter. The difficulties will argue for themselves...

Little did Churchill realize how truculently those difficulties would argue. The data about tides were abundant and precise. The tide tables could be refined to the hour and the minute. Far too little was known, however, about waves. The beach landings, the handling of thousands of ships, great and small, the construction of artificial harbours, and the visceral discomfort of tens of thousands of men depended on the behaviour of waves in this particular crooked elbow of the coastline of France.

Throughout the millennia, from the first coracle which put to sea to the giant liners, men had been at the mercy of the waves. Seagoing nations had built empires and opened up new continents by riding the waves. Generations of seamen had battled their way round Cape Horn or ventured the Northwest Passage, heaving and tossing or, like Coleridge's 'Ancient Mariner', had languished in the waveless doldrums. Waves had sculptured the shorelines of the land surface. An enormous amount of literature had been built up about waves, but it was largely descriptive and speculative with, as the emergency of the Normandy landings showed, surprisingly little scientific content.

A crash programme of research was needed to inform the planners. Scientists, with blackened faces and wearing commando uniforms, were smuggled ashore by night on the German-held beaches to make measurements of the configurations of the coastline and to take samples of the soil and of other factors which could influence the character of local waves. From those studies, made under the muzzles of the shore batteries of the West Wall, data were assembled rapidly and, with help from prior theoretical work by Harold Jeffreys, Harald Sverdrup, and Walter Munk, it was possible to relate them to the weather maps and so forecast the height and periodicity of the waves. The weather forecasts about gales which create storms (i.e. wave turbulence) required the postponement of D-Day for twenty-four hours, but it was the inshore wave estimations which presented the engineers with their problems of constructing the artificial harbours and quays, rather than Churchill's concern that they 'must float up and down with the tide'.

This salutary experience was a reminder that the oceans, embracing seven-tenths of the Earth's surface, were a neglected subject in science. One of the immediate results was the setting up, by royal charter, of the British National Institute of Oceanography and a flurry of upgrading, endowing, and integrating of institutes involved in this kind of work. The Inter-Governmental Oceanographic Commission was set up, with its headquarters in the Unesco Office of Oceanography, in Paris. Oceanography played a major role in the International Geophysical Year and subsequently in the Indian Ocean Project sponsored by Unesco, with the cooperation of survey ships of more than twenty nations.

After the Second World War, oceanography was thus promoted in the hierarchy of the sciences. The long reluctance to recognize it was considerably due to nomenclature – '-graphy' implies a descriptive discipline rather than fundamental research; it suggests a utilitarian function of charts to get ships from here to there. Although such problems, with emphasis on navigation, had been a main preoccupation of the Royal Society of London when Charles II founded it, the status of such 'heads of inquiry' was diminished by the growing specialization of the exact sciences. The Russians called the science 'oceanology' ('*logos*', meaning 'discussion') which gave it learned-society status in the Academy.

In past practice, oceanography had been a composite of four sciences: it included the physical study of current and wave movements; the geological study of the form of the ocean basins and their sediments; the chemical study of sea water and its dissolved substances; and the biological study of plant and animal life in the sea. The last suggests a rather artificial distinction *vis-à-vis* marine biology, which is defined as being concerned with the form, functions, and life histories of the plants and animals, whereas biological oceanography is considered to be concerned with the interrelationship between the plants and animals and the waters, currents, and sediments of the oceans. This, increasingly in the future, will be one of the most important aspects of oceanography because, with the multiplication of the world's population, the need for cultivating the resources of the sea becomes paramount.

Seven-tenths of the globe is covered by sea water but, considering the food which it produces, it is largely undeveloped. As far as the sea is concerned, we are still at the caveman stage – we hunt our fish; we do not husband them; nor do we cultivate the sea pastures. The total amount of seafood caught by all the nations of the world in 1963 was less than 30 million tons. The protein needed for the world's population could be supplied by doubling that catch.

The liquid mine

The ocean is also a liquid mine. It contains in solution, in suspension, or in deposition the Earth's elements, including scarce as well as abundant minerals. In this context, 'ocean' is used in the singular because, although we talk of the 'seven seas', there are no boundaries which prevent the transmigration of the waters, which are continually on the move, horizontally and vertically, through the movement of the deep currents and the surface driving of the winds. Although, around the coasts where fresh water comes in from the land, there may be a layer of fresh water less dense than the brine, and therefore floating on it, that too is churned up by the action of the waves.

For every 1,000 parts by weight, there are 965·1 of water and

34·9 of salts. The water is composed of 108 parts of hydrogen and 857·1 parts of oxygen. There is also a significant quantity of oxygen present as dissolved gas. Of the major constituents, 99·99 per cent of the salts are the cations of sodium, potassium, magnesium, calcium, and strontium; and the anions of sulphate, chloride, bromide, fluoride, carbonic acid, and boric acid.

The distribution of the elements varies with depth. For instance, just as land plants need nitrogen, phosphorus, calcium, and magnesium, so marine plants need phosphorus, nitrogen, and silicon. In the upper layers of the ocean, where the sun's rays penetrate and produce photosynthesis (the process by which plants capture the Sun's energy and use it to convert the nutrient elements into organic material), these three marine-growth elements are in much lower proportions than they are in the deeper layers. Below the level of photosynthesis they undergo increasing concentration because what has been converted higher up is released farther down by the dissolving of the sinking organic debris.

This is abetted by the metabolism of the bacteria and animals living in the deeps. There is also local supplementation. For instance, the phosphates of the prairies of the United States and Canada, which go into wheat crops and then into the Londoner's bread, go out with the sewage to feed the plaice (flounder) in the North Sea. And there are circumstances in which marked imbalances can occur. There can be oxygen depletion which can reach a point where all the dissolved oxygen (as distinct from the O of H_2O) is used up. This occurs when the amount of decaying material sinking from above exceeds the supply of oxygen. In those conditions bacteria types which obtain their oxygen supply by reducing sulphate ions multiply through lack of competition and produce the gas hydrogen sulphide, which is lethal to other forms of life. So, apart from the physical transfer of elements, there is a kind of biological escalator as well, the nature and range of which will have to be more systematically understood if the seas are to be turned into submarine pastures and ranges for the sea creatures we still have to domesticate.

Minerals are disposed in the ocean by various mechanisms. They can come from the mantle through volcanic vents and fissures. They can come from meteorites, for which the ocean offers a larger target than the land surface. They can be deposited in the form of cosmic dust. Mainly, however, they come from the crustal rocks of the land surface.

The weathering of rocks, the scouring of the rains, the action of the winds, and the 'open-cast mining' of the streams and rivers, carving their way from their watersheds to their respective oceans, contribute to the mineral deposits of the ocean. The waves themselves are a form of hydraulic mining, undercutting the cliffs, with their mineral formations, and surf grinding the hardest rocks.

Most of the continental shelves are covered with mud or mixtures of mud and sand and ooze, the latter formed by organic matter. Beyond the continental shelves are the continental slopes, like the slip edge of a slag heap. As on the shelves, the depth of sediment on the slopes may be as much as 20,000 feet. Sometimes the actions of the turbidity currents start an avalanche, and the unconsolidated sediments tumble deeper than the Grand Canyon, apparently maintained by turbidity currents which have the force of submarine rivers. Therefore, at sea, there is a continual transfer of land minerals. Some of those may be transported as pebbles, like the diamonds which can be dived for off the coast of western Africa, or like the gold nuggets which are scoured out of the alluvial or glacial deposits (like the placer mining of the Klondikers) and carried out to sea by the rivers.

Prospectors for the ocean deposits have now learned to look for other forms of concentrations. They can even find biological assistants. Certain sea creatures extract elements from sea water and concentrate them in various parts of their bodies. Vanadium, a rare element, is accumulated by certain holothurians ('sea cucumbers') and tunicates ('sea potatoes'). The proportions are remarkable – 50,000 more vanadium, by weight, than in sea water. Oysters effect a 200-fold concentration of copper. The bones of some fishes contain appreciable quantities of zinc, copper, tin, nickel, and silver. Sea snails also 'pack a load' of bromine but their help can probably be dispensed with because it is relatively easy to extract bromine from the sea water itself by first acidifying it and then chlorinating it. The liberated bromine is simply blown out by a current of air, to finish up as a constituent of ethyl gasoline. Similarly, magnesium is precipitated from sea water when it is made alkaline. The precipitate ('milk of magnesia' of the pharmacy) is filtered from the water. It can be converted into either magnesium carbonate or magnesium chloride from which metallic magnesium can be extracted by electrolysis.

Other minerals are deposited in bulk. For example, phosphorite, economically valuable as a source for phosphate fertilizers, occurs in relatively shallow waters (sometimes less than 300 feet deep) as flat slabs and irregular masses along the coasts of Australia, Japan, Spain, South Africa, the west coast of South America, and both coasts of the United States.

Nodular riches

By conservative estimates the oceanic waters contain 15,000,000,000 tons of copper, 7,000,000,000,000 tons of boron, 15,000,000,000 tons of manganese, 20,000,000,000 tons of uranium, 500,000,000 tons of silver, and 10,000,000 tons of gold. But sea water is a very 'lean ore'. To get even minimal amounts would mean chemically processing millions of tons of sea water. The sea, however, by its own alchemy and by the leisurely processes of thousands and millions of years, has already made substantial conversions.

They take the form of the so-called manganese nodules which incorporate other important metals. These were first brought to the surface in the 1870s by the British oceanographic vessel *Challenger*, which dredged them up from the deep parts of the Atlantic, Pacific and Indian oceans. From the turn of the century, when the *Albatross* expedition found that the nodules covered an area of the eastern Pacific larger than the United States, little deep-sea dredging for minerals was done until International Geophysical Year. The findings, around the world, showed that those nodules were, in composition and extent, a major source of economic minerals.

Manganese nodules have a wide variety of shapes and sizes. Generally they range from one to nine inches in diameter, but they can occasionally be enormous. One weighed 1,700 pounds. They are built up by a series of accretions – thin deposits like the layers of an onion. They need something to form a nucleus – a bit of clay, a basalt pebble, or even a shark's tooth. The manganese enters the sea by rivers, from springs in the ocean floor, or by the decomposition of submarine volcanic deposits.

The manganese ions react with oxygen and precipitate as molecules of manganese dioxide which clump together and migrate to the depths. The colloidal particles have an affinity for iron, which behaves similarly. As they filter down through the ocean they attract ions of nickel, cobalt, and other minerals. They are electrically charged and are attracted to hard-surfaced, electrically conductive materials on the ocean floor. They become like inorganic limpets. They go on growing by attraction of ions from the water. It is a very slow process; the quickest rate (estimated by radioactive dating) is reckoned at one millimetre per thousand years. To grow they need contact with the sea water, and if the rate of sedimentation is too high, they get smothered and cease to grow.

Some deposits yield as high as 80 per cent manganese dioxide, but samples from all the oceans average 32 per cent manganese dioxide, 22 per cent iron oxides, 19 per cent silicon dioxide, with smaller quantities of aluminium, calcium, magnesium, nickel, copper, cobalt, zinc, and molybdenum.

From an economic standpoint, these nodule deposits yield workable quantities of valuable minerals. For example, nodules rich in nickel and copper predominate in an area of about 14 million square miles in the south-eastern Pacific, where sea-floor photographs have indicated about 200,000,000,000 tons of such nodules. The mid-Pacific rise, which lies just west of Hawaii and includes the Society Islands, is rich in nodules which have a high cobalt content (apart from a third, by weight, of manganese). These deposits are abundant – perhaps five pounds per square foot of sea bottom over an area of 4 million square miles, and at a reasonable depth of 5,000 feet.

About 40 million square miles of ocean floor is covered with red clay, 300 feet thick. It consists of about 50 per cent silica, 20 per cent aluminium oxide, 13 per cent iron oxide, 7 per cent calcium carbonate, 3 per cent magnesium carbonate, and 6 per cent water. The proportions of manganese, nickel, cobalt, copper, and vanadium are of a different order, but in the immensities of the depositions the clay holds enough aluminium and copper to last a million years at the present world rate of consumption. The copper is ten times more abundant in the unconsolidated clay than it is in the igneous rocks from which it has to be laboriously mined on land.

The world needs cement. Globigerina ooze, consisting mainly of the skeletons of protozoa, covers 50 million square miles of the ocean floor. It rates as high as 95 per cent calcium, as rich as the limestone rocks from which we make cement. The estimate of these submarine calcium deposits, even if only 10 per cent were cement grade, gives 100,000,000,000 tons.

Oceanographers themselves have shown how the wealth of the seas could be 'mined'. For over a century they have used a drag dredge to collect their samples even from depths as great as 30,000 feet. It is just a bucket, hauled along the ocean floor. But modern possibilities can be much more wholesale than that. A hydraulic dredge, like a giant vacuum cleaner, can remove the nodules from the sea bottom without disturbing the floor itself; television cameras can locate the deposits and supervise the operations. The technology is not forbidding and the capital costs should work out at less, with only water or overlay, than the massive land operations for mining materials.[1]

Staking claims for this submerged Klondike raises profound issues of international law. Nations have jealously insisted on their offshore rights (for fishing and for use of tidal oil lands), but even if they extended their claims to the whole continental shelf, adjoining and extending their terrestrial domains, they could not claim sovereignty over the deep-sea deposits and, as has been pointed out, the nodule riches are to be found away from land-deposited sediments. So the three-mile or twelve-mile limit of international dispute would not apply. New laws of the sea would require some regulatory body to allocate ocean mining rights. The United Nations, by becoming such a body and taking

1. See John L. Mero, 'Minerals on the Ocean Floor', *Scientific American*, December 1960, pp. 64–72.

even minimum royalties, could, over the next half century (when ocean mining will have developed on a vast scale), finance its entire operations.

The sovereignty of the seas

The United Nations' sovereignty and responsibilities for the ocean should include more than minerals. If the sea is to be cultivated and its creatures domesticated into the world's food economy, no one country would be expected, nor entitled, to exploit the submarine rangelands and pastures. For example, British marine biologists showed long ago that it is perfectly feasible to have land-based nurseries for fish, e.g. plaice and sole, and to rear the fingerlings through the stages in which, in nature, they are the prey of countless enemies. When they are better capable of looking after themselves they can be restored to their normal feeding grounds. But such fishing grounds are open to all comers, and only some international organization should be expected to finance such an enterprise (on a really big scale) to supply the common need for increased seafood. Another proposal is that there should be a tidying up of the sea bottom. The principal competitors of the food fish for the nutrients of the sea are the starfish. If they could be weeded out (and converted into poultry food) the edible fish would better flourish. Again, that is an international matter as well as a common opportunity. Yet another suggestion is that there should be systematic development of the Southern Ocean, surrounding the continent of Antarctica, which has already been taken out of colonial competition by international agreements. This is the great region of summer plankton growth, which is greater as a store of the Sun's energy photosynthesized than are the wheat prairies and pastures of the land.

The creature which lives abundantly in those sea pastures of the south is the krill, a planktonic shrimp, about two inches long. It is found in immense, densely packed shoals and is the food of the giants. The whales which it nourishes have the all-nature record for growth. From a microscopic ovum, the embryonic blue whale grows to twenty-three feet at birth. The mother whale's sole diet is krill and, on a diet of krill, the offspring is sixty-five feet long within two years. Because of anarchic whale fishing, the whale stocks have been dangerously reduced. But massacring the whales in order to get the krill, which they have converted into blubber, is like Charles Lamb's story of the Chinese who burned down his house every time he cooked his pork. Sir Alister Hardy proposed that there should be mechanical whales – nuclear-powered submarines, perhaps – which would swallow the krill just as the natural sea mammoth does, and convert it into food forms for the needs of hungry people. Again this ought to be the job for United Nations Incorporated.

It has always been in the common interest of nations to share information about the high seas, and so the scientists have been the beneficiaries of admiralties and naval departments.

The details, however, are being increasingly filled in because of the availability, range, and precision of new instruments. Echo sounding was a by-product of war. Pierre Langevin, the French physicist, in the First World War had proposed using sound echoes as a means of detecting submarines. The reception of a reflected sound signal bouncing back from some submerged object can give its location, and if the signals are refined enough and the receptions discrete enough they can build up an image (though rather blurred) of the object. Sound pulses thus do what electromagnetic pulses do in radar. In naval parlance, echo-sounding systems are known as 'asdic' or 'sonar'.

The difficulties of perfecting the echo-sounding systems have arisen from the refraction, scattering, and absorption of sound. The refraction is due to the bending of the sound waves, the speed at which they travel being varied by water pressure, salinity, and temperature. Radar has the advantage of the constancy of the speed of light and, therefore, distances can be neatly calculated as a factor of the absolute time it takes for the signal to go out and come back. Echo sounding requires much more complicated calculations, although, conversely, it builds up a great deal of information about the hidden factors which affect it. Scattering of sound, for instance, results mainly from the living organisms which migrate vertically, particularly at night. Echo sounders have revealed that what is called a 'deep scattering layer' is due to shoals of small fish, particularly the luminous myctophids or lantern fish.

Absorption interference is due to the existence of magnesium sulphate in the sea water which acts as an acoustic trap, particularly for the high frequencies which could give precision to the picture drawing.

Hydrophones can give another kind of sound picture of the depths of the sea. It has been called 'the silent world beneath the sea', but the recordings have revealed it as a jungle of sound, alive with the chattering and bellowing of sea creatures, from the shrimp chorus near the surface to mysterious animals in the depths which have never been seen.

What has emerged from all this is a geography of the sea bottom, greatly enhanced by the International Geophysical Year of 1957–9 and the Indian Ocean Project of 1959–65.

Birthplace of the monsoons

Two thousand five hundred years ago, Darius I, the Persian emperor (522–486 B.C.), ordered a fleet to be built at the junction of the Kabul River and the Indus, in the Indian subcontinent. He appointed Scylax of Caryanda to command the fleet and ordered him to sail down the Indus and survey the sea route to Egypt. Scylax took thirty months to fulfill his mission, and the culmination was the digging of a forerunner of the Suez Canal, between the Red Sea and the Nile. 'Its length,' wrote Herodotus, 'is four days' journey, and the width such as to admit of two triremes [galleys with three banks of oars] being rowed along it abreast.'

Thus, by history and engineering, the Mediterranean was linked with the Indian Ocean. But until Vasco da Gama (*c.* 1460–1524) rounded the Cape of Good Hope in 1498 and sailed to India, Europeans never ventured into the Indian Ocean, which Arab traders had plied for centuries.

It is an extraordinary ocean, bounded on the north by Pakistan and India, on the west by the Arabian peninsula and Africa, and on the east by the Malay peninsula, Indonesia, and Australia. Its extensions are the Red Sea and the Persian Gulf. It is 28,400,000 square miles in extent.

It is extraordinary not only oceanographically, but meteorologically. In no other part of the world does the regular half-yearly alternation of all the factors in weather-making play such a conspicuous part. It is the birthplace of the monsoons. From October-November to March-April north-east winds prevail in the north latitudes and north-west winds in the south latitude. From May-June to September-October south-west and south-east winds prevail. The south-west monsoon carries the rains and and the north-east monsoon is predominantly dry. Southward from the latitude of the Seychelles the south-east trade wind prevails throughout the year, and south of 30° S. the west winds are particularly strong.

This regular routine of winds governed the voyages (and the fates) of ships to the Orient. They came and went on the alternating winds. These winds carried the Arab dhows to extend trade, and Islam, to the Indonesian archipelago probably a thousand years ago, and they carried the ships of the conquering empires, Portuguese, British, Dutch, and French.

And with the winds went the surface currents. In the northern part of the Indian Ocean these currents change with the monsoon. During the north-east monsoon the currents flow towards the south-west and west, and between the easterly current maintained by the monsoon and the westerly current maintained by the south-easterly trade wind a strong countercurrent flows east at about 10° S. The south-west monsoon acts as a continuation of the south-east trade wind in the Southern Hemisphere, turning in a south-westerly direction when it crosses the Equator. A strong northerly flow is found off the coast of Somaliland and the Arabian peninsula, where the near-shore temperature is lowered. In the southern part a counterclockwise circulation prevails, with the easterly currents representing the Antarctic circumpolar flow.

Thus the Indian Ocean is a weather cauldron, continually stirred and boiling up into the violent cyclones which have made the Indian Ocean a dangerous place even for modern steamships.

In the latter part of the nineteenth century, H.M.S. *Challenger* and the German ships *Gazelle* and *Valdivia* made soundings of the ocean bed. British cable ships linking up the Empire by telegraph also acquired a lot of practical knowledge about the bottom. But the systematic investigations really began with the extensive cruises of *Discovery II* which began in 1929 in the western part of the ocean. Many ships of many nations continued the exploration, but the present era began with the historic voyages after the Second World War of the Swedish *Albatross* and the Danish *Galathea*, the first in the equatorial regions, and the latter in the abyssal waters between South Africa and Indonesia.

Curiosity about this unpredictable ocean was further excited by the discovery of the coelacanth. This fish is a 'living fossil'. It existed 350,000,000 years ago and was supposed to have become extinct 60,000,000 years ago. The *Coelacanthini* and its related order the *Rhipidistia* were the collateral branches of the creatures that populated the Earth. They 'experimented' as amphibians. The rhipidistians opted for the land and the coelacanths remained aquatic.

A living coelacanth (*Latimeria chalumnae*) was trawled up from 240 feet depth off East London, South Africa, in 1938, and a second was caught by a native fisherman with a line, again from 240 feet, near the Comoro Islands in December 1952. Both specimens were decayed and mutilated before the experts were able properly to examine them. In 1953 and 1954 five more specimens were captured alive and preserved. They were almost unchanged in form, since 60,000,000 years ago. They had limb-like fins which could be used for crawling about on the sea bottom. This discovery raised all kinds of exciting speculation about the aquatic museum which might exist in the depths of the Indian Ocean.

9 Crustal Movements

The rifts

As a follow-up of the International Geophysical Year, the Indian Ocean Project was promoted by the Inter-Governmental Oceanographic Commission, sponsored by Unesco. More than two thousand scientists from over twenty countries operating from thirty research ships concentrated their investigation on this particular region between 1959 and 1965. Linked with Unesco were the Food and Agriculture Organization, the World Meteorological Organization, the United Nations Expanded Programme of Technical Assistance, and the United Nations Special Fund. The investigations were supported by a computer based in Bombay which processed 20,000 weather observations an hour.

Two of the strange phenomena – the hottest, saltiest water in the sea and the coldest surface water in the tropics – were subjects of special study. The U.S. ships *Atlantis I* in 1958 and *Atlantis II* in 1963 had found abnormally hot, salty water near the bottom of the Red Sea near 21° N. The British ship *Discovery* on 11 September 1964 anchored a radar buoy in 1,400 feet of water and sent down bottles. The radar signalled a sudden change of temperature but the scientists were surprised when the thermometers in the bottles were recovered. The normal Red Sea water temperature was around 22° C. (71·6° F.), but at depth the thermometer registered 58° C. (136·4°F.). The salinity was equally surprising. When water was being drawn from bottles that had been near the bottom it proved to be an unusual liquid. It poured with difficulty and any spilled on the deck immediately dried leaving a thick, white patch of crystals. It was so salty that the usual instruments could not measure it. When experiments were carried out, the salinity was 271 parts per thousand compared with the average Red Sea salinity of 40 parts per thousand. This study was followed up with a second survey by *Atlantis II* and by the German ship *Meteor*. The salinity was found to be comparable with that of the Dead Sea, the saltiest water on Earth.

Most interesting information was obtained by sediment cores from the bottom and the sides of the deeps. The material when first brought to the surface is tarry black in appearance and is principally composed of iron oxides (magnetic), anhydrite ($CaSO_4$), and amorphous silica (SiO_2nH_2O). The X-ray patterns also show small amounts of sphalerite (ZnS). These findings raised basic questions concerning (1) the origin of these hot brines, (2) the high concentration of iron, manganese, and silica in the aqueous phase, and (3) the chemical relationship between the hot waters and the deposition of silica, anhydrite, and heavy metals.

In 1957 a Soviet research vessel, en route from Ceylon to the Gulf of Aden, encountered a mass of dead fish floating over 60,000 square miles, a tonnage equal to the world's commercial fish catch for an entire year. One assumption was that this catastrophe had been caused by the upwelling of currents of water with little free oxygen, but the investigation of the Somali Current gave a different explanation. This current was found to be flowing north at a speed of seven knots. At one point within ten degrees of the Equator, surface water temperature was found to be only 13° C. (55·4° F.), compared to the normal 22° C. (71·6° F.). This cold water coincided with the 200-mile-long zone of high fish mortality.

Systematic charting was made of the ocean floor by ships on a carefully planned grid system, so that the picture given by the echo soundings was like the overlapping aerial survey mosaics produced in land-surface mapping by aircraft. A north–south ridge of submarine mountains, running 2,250 miles from Malaya to Australia, was discovered by the Soviet survey vessel *Vityaz*. The findings of other ships, using gravity and magnetic surveys, found that the African continental structure continued for more than 200 miles, as far as the Seychelles Islands. They also charted the Mid-Indian Ocean Rift and showed that it was an extension of the Mid-Atlantic Rift, which extends, as has been mentioned, from the North Polar seas down the length of the

North and South Atlantic oceans and around the Cape of Good Hope. It was now found to bend northward again as far as the Red Sea.

This work by the oceanographers linked up dramatically with the work of other scientists of twelve countries studying the African Rift. The African Rift runs 2,000 miles south from the Red Sea through Ethiopia, Kenya, Uganda, Tanzania, and Mozambique and extends north to the Palestine Rift, in which lies the Dead Sea and the Jordan Valley. It has been formed in the last 15,000,000 years as a result of the great upward thrust that has rent the Earth's surface in East Africa. The rift floor is twenty miles wide in some places and more than 1,500 feet below the tops of the escarpments lining it.

A typical cross section of the Mid-Ocean Rift has a profile strikingly similar to that of the African Rift. Indeed, the continental formation appears to be a landward extension of the Carlsberg Rift, of the Indian Ocean, which comes ashore in East Africa.

The Indian Ocean Project was not only a study of the topography of the 28,400,000 square miles of the planet's surface drowned beneath the sea, or of its currents and marine biology; it included the study of the atmosphere, with aircraft cooperating with the ships. It was, as I.G.Y. had been, a cooperation of many disciplines, and it had a close affinity with that other impressive feature of I.G.Y., the investigation of Antarctica.

Antarctica

Before the International Geophysical Year, very little was known about the continental landscapes below the ice of Antarctica, and what was known was almost entirely derived from observations around the rim of the continent. Soundings in the surrounding ocean indicated that the Antarctic continental shelf lay at a depth of about 1,300 feet compared with 30 feet for the shelves of all the other continents. This was impressive (in both senses of the word) evidence that the enormous weight of overlying ice was pushing down the Earth's crust. In other words, the sial rafts supporting the terrestrial rocks of the continent were pressing into the plastic sima, like a body into a mattress. What rock surface was visible consisted of outcroppings in the low coastal areas, a few cliffs, and the nunataks, mountain peaks projecting through the ice of the high plateau. The great indents of the Ross Sea and the Filchner Ice Shelf on opposite sides of the continent suggested that a huge trough divided Antarctica into two parts.

One of the principal objects of the I.G.Y. was to obtain more information about the land below the ice. One of the principal means, apart from traditional exploration, was the use of seismic 'shooting'. The reflected and refracted waves from man-made explosions could give the depth of the surface of the underlying rock and provide information about the physical nature of the various layers of the Earth's crust. Instruments for measuring gravity were used in an area where the centrifugal force which offsets gravity decreases to zero at the geographic pole. By such measurements it was found that the underlying crust in the Byrd Station area had been warped to about 3,000 feet and that the crust would rebound if the ice load were to melt. By such remote detection (resembling echo sounding of the oceans) the topography of the land, buried under more than 6,600 feet of ice, is being mapped.

The amount of ice covering the 4,400,000 square miles of the Antarctic continent (excluding the floating ice shelves) has been calculated. Since the area of the ocean is about thirty-two times the area of Antarctica, if all of the ice melted the level of the oceans would rise about 200 feet, if the continent did not rise up or the ocean floor sink when the melt was transferred to the seas. If the ocean floor were to sink (like the lowering of the bulge in the mattress when the dent flattens out), the estimated rise in sea level would be reduced to 130 feet. If this were to happen, the picture of the world's coastline would be drastically changed. Remember that the melting of the North Polar ice shelves, which are floating ice, would not raise the level of the seas because the displacement of the floating ice is equal to the water which would be melting. Land-held ice, as in Antarctica and in Greenland, is a different proposition – it would be an addition to the water of the ocean.

The question immediately arises as to whether the quantity of ice on Antarctica is increasing or decreasing and how fast. Every year an average of one to two feet of snow falls on Antarctica. Most of it remains to be gradually compressed into a layer of ice with an average thickness of ten to twenty centimetres which weighs one to two million million tons. The recent surveys show that the surface slopes towards the coast all over the continent and, therefore, the ice flows everywhere towards the sea. (However slow the movement, it persists.) The ice water discharges into the ocean from the glaciers. The problem is to establish whether the outflow is in balance with the annual accumulation over the continent. There are figures for the outward creep of smooth areas of ice from the coastal mountains and data are being accumulated on the velocity and thickness of the wider and more rapidly flowing glaciers. Estimates of the ice budget range from a gain of 1,320,000 million tons to a loss of 410,000 million tons.

The estimated rise of the world's oceans was about 1·2 millimetres per year during the first part of the 20th century. In the general warming of the climate (which will be discussed later) the most vulnerable part of Antarctica would be the eight square miles of the ice shelves, which, being floating ice, would not affect the sea level, and inland ice below sea level, the melting of which would raise the sea level from seven to twenty feet.

When the I.G.Y. ended in 1959, a Special Committee for Antarctic Research (S.C.A.R.) was set up to coordinate the scientific work of the nine nations which had agreed to maintain some forty stations on the continent and islands. These nations are Argentina, Australia, Chile, France, New Zealand, the Republic of South Africa, U.S.S.R., United Kingdom, and the U.S. These countries all had footholds and territorial claims on the empty continent before it became inhabited by international scientists.

The scientific cooperation developed during I.G.Y. led to one of the most hopeful of diplomatic agreements. In December 1959, a treaty was signed and immediately ratified by all parties who 'froze' all territorial claims. The treaty provided for the demilitarization of all national bases on the continent, with full unilateral rights of inspection. It banned nuclear explosions and the dumping of radioactive waste, and set up machinery for the complete exchange of all scientific information.

The continent became a great scientific laboratory. The continuing researches include: aurora and airglow, biology and medicine, cosmic rays, geodesy and cartography, geology, geomagnetism, glaciology, meteorology, oceanography, seismology, transverses, and upper-atmosphere physics.

Lost continent refound

There are strong possibilities that Antarctica is the lost continent of Gondwanaland, a missing part of the jigsaw puzzle of Wegener's Theory of Continental Drift.

In 1885, even before Wegener, Eduard Suess, the Austrian geologist, had suggested that there must have existed a gigantic continent which he called 'Gondwana'. He invoked it to explain how the sedimentary rocks of India, Madagascar, South Africa, New Zealand, Australia, and South America had so much in common. They must, he argued, have been all one at some stage. This suggestion would accord with the Wegener Theory, but when all the bits have been tucked in together there is a hole – a missing part – in what is now the Indian Ocean. Antarctica fits rather nicely, if, as has been suggested,[1] the Scotia arc is treated as a hinge to swing it round South Africa.

The first explorers, who were in fact the first human beings, to set foot on Antarctica were amazed to find evidence that plants and animals had once flourished there. They found trunks of petrified trees, the imprints of leaves on rocks, and coal seams which demanded for their existence a one-time abundance of tropical forests. The records are all there of a succession of plant life from first marine forms to quite late ancestors of our present forest trees. Animal remains exist as fossils. From radioactive dating Soviet geologists have established outcrops in East Antarctica at about 1,450 million years ago. Near the Wilkes Station, also on the eastern part of the continent, the rocks are 400 million years younger and consist of intrusive bodies of granite forced up from the interior of the Earth. In the glacial debris found at the bottom of the Beardmore Glacier (which must have come from the highlands of the interior) have been

1. See p. 139.

found fossils of a coral-like marine organism (which can also be found in Australia) that show life 600 million years ago.

During the I.G.Y. there were found Devonian sandstones and shales in which were trapped a great variety of invertebrates. These were evidence that there must have existed shallow, warm waters with sandy and muddy bottoms in which the creatures had existed. The specimens were different from those found in the Northern Hemisphere, but it is clear that on land which now lies within 300 miles of what is now the South Pole, there once were creatures identical with the South African and South American fauna.

Later, a mere 250 million years ago, the evidence of the rocks shows an abundance of vegetation. Some of the coal beds are thirteen feet thick. In the formations are trapped petrified tree trunks twenty-four feet in length and two feet in diameter with fossil-leaf evidence which suggests lush, green vegetation in a humid, swamp environment. The Antarctic coals are of high quality: anthracite, which indicates that the primeval forest and swamps were buried under an enormous thickness of sediment. Those deposits and sedimentary rocks were invaded 180 million years ago by hot, molten material from below which baked the deposits.

There is evidence that throughout the Tertiary Period, which embraces most of the earlier part of the past 100 million years, West Antarctica was involved in the great mountain upheaval which created the Andean chain of South America.

During the Indian Ocean Project, the Soviet geophysicists reported that no evidence had been found to bear out any theory of a 'collapsed' (i.e., sunken) continent in the Indian Ocean basin, but the mapping showed that the sea bottom, with its ridge systems, was markedly contrasted with that of the Pacific, which is regarded as a primary ocean bed. The characteristics of the Indian Ocean bottom might suggest a 'removed' continent – like a table scratched by the sliding of a tray. This could have been the result of the breakup of the land mass and the migration of the Antarctic continent from tropical or temperate regions towards the South Geographic Pole, the end of the Earth's axis.

The I.G.Y. evidence from Antarctica shows that the South Magnetic Pole has also 'wandered' like the North Magnetic Pole, and examination of the continent's magnetic crystals ought to show the course which the moving land mass took.

The Arctic

At the other end of the Earth's axis, the earth sciences and the life sciences combined to throw light on the Americans who were in the Western Hemisphere before Columbus.

Anthropologists generally agree that the first immigrants came by way of a land bridge linking the north-east corner of Asia to the north-west corner of the United States across the Bering Strait. This was backed by marine soundings which showed that the strait was not only narrow (fifty-six miles wide) but also shallow. A lowering of the present sea level by a hundred feet would change the strait into an isthmus across which entry could have been made on foot.

Twentieth-century scientists, however, began to question this plausible explanation because the migration from Asia to America, and vice versa, was not confined to enterprising human beings, but included movement of animals and plants. This was scarcely consistent with the idea of a comparatively narrow causeway. Geologists would now maintain that the Old and New Worlds were joined by a broad land mass which is now submerged on both sides of the Bering Strait.

South of the strait is the Bering Sea. Its floor is one of the flattest and smoothest stretches of terrain on the entire globe. With a slope of no more than three or four inches to the mile, it reaches southward to a line that runs from Unimak Pass in the Aleutians to Cape Navarin on the Asian shore. Along this line – the edge of the continental shelf – the sea floor plunges steeply from a depth of about 450 feet down 15,000 feet to the bottom of the ocean. The floor of the Chukchi Sea, north of the Bering Strait, is not quite so smooth; the depth varies from 120 to 180 feet, and irregularities of the terrain bring rock shoals upward to depths of only 45 feet and lift the great granite outcrops of

Wrangel and Herald Islands above the surface of the sea. Along a line that runs several hundred miles north of the Bering Strait from Point Barrow in Alaska to Severnaya Zemlya off Siberia, the sea floor plunges over the northern edge of the continental shelf to the bottom of the Arctic Ocean.

Soundings of the Bering and Chukchi Seas thus reveal a vast plain that is not deeply submerged. At its widest the plain reaches 1,300 miles north and south, 600 miles wider than the north–south distance across Alaska along the Canadian border. The granite islands that rise above the water testify that the plain is made of the same rocks as the continents.[2]

During most of the 50-million-year duration of the Tertiary Period there was thus a great tundra plain to cross, which helps to explain the paleontological evidence of the large and small mammals which moved from Asia to America. The submergence of the land should have interrupted this migration, and there is no evidence that the land rose up again during the million-year Pleistocene, during which early man emerged. Nevertheless, fossil evidence shows that numerous animals, large and small, crossed from Asia to North America during the Pleistocene. Rodents certainly crossed and spread southward through North America, but not into South America. Later came the larger mammals: the mastodon and mammoth, musk oxen, bison, moose, elk, mountain sheep and goats, camels, foxes, bears, wolves, and horses. The horses flourished and then died out in North America; the genus was not seen again in the New World until the *conquistadores* brought their animals across the Atlantic.

The Pleistocene Epoch includes the great Ice Ages when the land ice locked up great volumes of water, i.e. water extracted from the sea by evaporation and deposited by atmospheric precipitation to become ice on the land. Studies of the Mississippi Gulf region have provided precise information about the course of the last great Pleistocene glaciation, the so-called Wisconsin Period. It has been determined that the Wisconsin Glacier reached its maximum 40,000 years ago when it had withdrawn

2. William G. Haag, 'The Bering Strait Land Bridge', *Scientific American*, January 1962, pp. 112–23.

enough water to lower the sea level by as much as 460 feet. This (without assuming a geological rise in the land) would have exposed the sea bed of the Bering Strait. With such a 460-foot drop of sea level, there must have been a dry land corridor between the continents 1,300 miles wide, and large animals moved freely across it for 80,000 years. Towards the end of the Wisconsin Period, men, following the animals which they hunted, must have crossed the bridge. Their remains on the bridge, however, are unlikely to be dredged up because after the seas rose again, with the melting of the ice covering the North American continent, sediments a hundred feet deep were deposited.

This evidence would put the first man in America at about 50,000 years ago, with the implication that the Bering 'drawbridge' was opened behind him. This does not explain the great diversity of American types. There must have been later migrations. The Eskimos were obviously latecomers because they are well-identified Mongols with a physical appearance which is indistinguishable from their neighbours, the Chukchi on the Asian side, with quite late characteristics such as the Mongolian eye fold. The widely different characteristics of the Indians in both North and South America suggest migrations across Asia from as far away as the Persian plateau, perhaps from the Near East and even Africa. For example, the hawk nose of so many of the Indian tribes is not Asian in the Far Eastern sense. It is more Aryan, like the Iranian types today. Other types of Indian are strongly suggestive of Egyptians (anthropometric measurements of the skulls of the Arizona Basket Makers and of the Coahuila cavemen agree uncannily with the cranium dimensions of the ancient Egyptians). The Pecos Indians have been described as 'pseudo-Negroid' and this refers not to the forced migration of the African slaves but to remotest history. Other American Indian types have been identified with the Dravidians of the Indus Valley – suggesting a common origin for the Indians of both hemispheres. The correlations have led to suggestions that there must have been a transatlantic *Kon-Tiki* which would allegedly explain the affinity between the calendar systems of the Nile and Central America. The explanation is more likely to lie in a 'long haul' eastward across the land mass of Asia. Those inherited traits of well-established genetic types, however, take a lot of reconciling with the geological ups and downs of the Bering land bridge. Indeed, the only excuse for this digression from the examination of the physical world into anthropology is to emphasize the need for a unity of science in which the various disciplines reinforce each other instead of being treated as divergent branches.

Prehistoric Boston Tea Party

Earlier reference was made to the 'unstitching' of the seams of the Earth's crust in the mid Atlantic – the longitudinal twin ridges stretching from Jan Mayen Island in the Arctic to Bouvet Island on the fringe of Antarctica, and sweeping round into the Indian Ocean. (Incidentally, the ridges have an interesting parallelism with the outlines of the European and African continents.) There are also transverse ridges like the Rio Grande–Walvis ridges, which, through Tristan da Cunha, link South America and Africa, or that which links Ascension Island and Africa, or that which crosses the north–south ridges in Iceland and links Greenland with Europe.

The rock men have joined forces with the volcano men and have shown that the rocks of the Atlantic islands tend to increase in age with increasing distance from the Mid-Atlantic Ridge. For instance, the oldest rocks of the Bahamas are 120,000,000 years old and those of the Azores only 20,000,000. Tristan da Cunha Island is no more than one million years old. Iceland is schizogenic; it has three areas of different ages, varying from 10,000,000 to 50,000,000 years.

All this raises fascinating questions as to when America cut adrift from Europe. When did the geological 'Boston Tea Party' occur?

There are fairly convincing relationships between Old Scotland and New Scotland (Nova Scotia). In Scotland, 350,000,000 years ago, there was a split, or fault, between the northern and southern parts. The upper part skidded 60 miles to the south-west along the line of what is now the Great Glen. In Nova Scotia, now

2,750 miles away, there is the Aspy Fault, which has precisely the same characteristics as the fault in Scotland and is part of the Cabot Fault system. This fault extends from Newfoundland to Boston. The two faults would be one if we apply Wegener's reconstruction. He also predicated a fault between Greenland and Ellesmere Island. The Geological Survey of Canada proved, posthumously, that he was right.

It is, however, easier to redraw the maps than to explain how Nature drew them in the remote past. The idea of continents on their sial rafts sliding apart on the sima is a simple one – so simple that earnest scientists disputed it – but what set the movement going?

Recent methods give an insight. We have talked about gravity, but there is such a thing as 'diminished gravity'. Over the abyssal trenches in the sea floor, such as are found in the Pacific, some of the largest deficiencies of gravity are recorded; some force there sucks the crust into the trenches and it is more powerful than the pull of gravity. There is a scientifically satisfying explanation: the interior of the Earth is in a state of extremely sluggish thermal convection, turning and churning the way water does in a kettle when it is on the boil. This usefully accounts also for the transfer of heat flowing from the Earth's interior through the mantle, the region between the core and the crust. Trenches, such as those in the Pacific, mark the places where the currents in the mantle descend again. This you can understand in the light of the example of the kettle, which lifts the lid and then sucks it back again; in the process marked by the trenches, the ocean floor is sucked in.

This is now a contemporary paradigm: convection currents in the mantle are generally accepted and generally confirmed by incidental intelligence. They flow very deep and move only a few centimetres a year so their actual behaviour is difficult to measure, but they are consistent with what one might call negative theory. We have seen how, on the principle of the body on the mattress, there is a slow recoil when the weight of glacial deposits is removed. This is an indication of the viscosity of the mantle material which would have to be 10,000 times more viscous to *preclude* convection. But the evidence for the down suck (the trenches) can be assumed from the manifest evidence of the upthrust (the volcanic arcs). In other words, we can see the lift of the kettle lid. Like the escape of the steam when the lid is pushed up by the convection forces, the flow of heat along the Mid-Atlantic-and-beyond ridges varies from two to eight times the temperatures observed on the continents and the rest of the ocean floor – except the deep trenches where, conversely, the heat falls to one-tenth of the average. In any get-together of contemporary oceanographers, the consensus would be that the ridges form where convection currents bubble up, and trenches form when they suck down.

From here on, common sense takes over: it is easy to believe that where convection currents rise and separate, the surface shell of the Earth's crust breaks by tension and a widening crack occurs: the laceration, mentioned earlier, which is filled up by the exudation of basalt lavas from the mantle. The land mass is thus given a thrust. The newer islands which were mentioned are the upgrowth of the 'scab' of this self-sealing wound. Secondary effects are the catastrophic earthquakes. Those are no more than the breaking of the surface material which is chilled and brittle and which cracks. This can explain the original break-up, multiple fracture of the continental land mass, but it is also consistent with the periodicity of mountain building and the random scattering of the continents. They would be pushed and swung around by the bubbles of convection, like the pieces of toast in boiling onion soup.

In this general process, the system of faulting becomes naïvely credible. The crust breaks and shifts. Then it heals like a badly set broken bone, with the parts permanently displaced – by sixty miles as in the case of the Scottish fault. Then the healed structure gets fractured again and the parts get translated, like the Aspy Fault in Nova Scotia. As the late Professor Wegener might have said, 'Q.E.D.'

One might add that the general pattern shows that the ocean floor is faulted at right angles to the Mid-Atlantic Ridge. These are ancient and inactive fractures, while the longitudinal fractures are still active, as indicated by the earthquakes along the San Andreas Fault in California, where the displacement –

mountain ranges snapped off and the parts separated – can be clearly seen from the air.[3]

3. J. Tuzo Wilson, 'Continental Drift', *Scientific American*, April 1963, pp. 86–100.

10 Climate

The bathtub whirl

A jaded question in broadcast quiz games is, 'Which way does the bath water whirl down the plug hole?' The answer has been in the textbooks since 1835 when Coriolis, a French engineer, described the force which bears his name. The motion of every particle on the rotating Earth, is governed by the rotation of the Earth. The movement is proportional to the latitude, so that it is maximum at the Equator and zero at the poles. The direction of the Coriolis Force is to the right (clockwise) in the Northern Hemisphere and to the left (counterclockwise) in the Southern Hemisphere. Water, in a bath or in the sea, consists of particles which are governed by this force. The effect on the oceans is part of the machinery which makes our globe the complicated engine which regulates (however irregular we may think it is) the day-to-day conditions of our lives: whether farmers are going to lose their crops; whether a hurricane is going to blast a certain area; whether the fishing fleets are going to catch fish; or whether a *Kon-Tiki* raft is going to drift to Easter Island. In other words, the movement of ocean currents – inextricably bound up with atmospheric currents, and hence with our weather – is basically dominated by this force. Cocksure oceanographers would tell you that, given systematic information, the oceans would reveal more about long-range weather forecasting than radiosonde balloons.

Maybe that is true. Short of that, however, the behaviour of those currents is of paramount importance. Benjamin Franklin realized that when, on his way to Britain to argue about the Stamp Act, he made observations of the Gulf Stream. Thor Heyerdahl, more romantically, when he launched his balsa raft into the Humboldt Current off Peru, had already convinced himself that it would be carried to Polynesia.

Without the breakwaters of the continents, the currents would spin uninterrupted round the globe like waters round the plug hole – clockwise in the north, counterclockwise in the south. Instead they form an extremely complicated system which, though picturesquely described in the past, has been scientifically studied (literally, in depth) and measured only in the second half of the twentieth century.

The tidal factors, influenced by the gravitational drag of the Moon, are familiar. There is the regular pattern of the ebb (seaward) and flood (landward) and this expresses itself along the confining coastlines as high water and low water. Tidal currents, however, exist in the open sea as well as in restricted channels. In deep water their effects are barely measurable, but on the continental shelves they are the predominant currents. Away from the interference of the coastline they change direction, and a drifting object completes a roughly circular path every twelve hours. The diameter of the circle is usually about four to five miles.

Wind currents are like breath blowing tea in a saucer, on a massive scale. If the wind is constant for any length of time, the friction between the moving atmosphere and the layers of ocean waters sets a strong flowing current in motion. It does not strictly follow the direction of the prevailing wind because it has to contend with the Coriolis Force, which deflects it, as would happen if the saucer were spinning. The surface layers set in motion by the winds act on the successive layers underneath, but at diminishing speeds and increasing deflection, until at a certain depth the current reverses, and, at a speed one twenty-third that of the wind-driven surface current, flows in the opposite direction.

There are, however, currents which are independent of the driving winds. These currents would justify the division of the geographically undivided ocean into 'the seven seas' because there are seven distinct circulatory systems: the Arctic, the Antarctic, the North Atlantic, the South Atlantic, the North Pacific, the South Pacific, and the Indian Ocean. These are dominated by intense currents such as the Kuroshio in the North Pacific, the Humboldt in the South Pacific, the Gulf Stream in the North Atlantic, and the Agulhas in the Indian Ocean. Such permanent currents so offset the powerful forces of the wind that very little water crosses the equator though there is some exchange at depth in the Atlantic.

Modern shipping of all kinds cooperates in the scientific evaluation of the ocean currents. Hundreds of ships every day report to the hydrographic institutes of the world – doing for ocean currents what the weather ships and weather stations do for atmospheric changes.

The methods of determination vary from the old familiar 'shipwrecked-sailor' device of the message in the bottle which, on recovery, can show how far it has drifted (but no details of its wanderings) to highly sophisticated electronic methods. A modern variant of the drifting bottle is the drogue which, attached to a surface light buoy or radar reflector, can be sunk to predetermined depth. In this way a ship can keep observational track of it. A further refinement of this is a self-contained instrument which, with a sound-generating apparatus and no surface connexion, can be sunk into the deep currents. It will keep on 'pipping' as it is carried along and its course can be charted by hydrophones. By this means currents can be measured at depths of 5,000 feet.

Sea water is a good electrical conductor and this property has been exploited in 'G.E.K.' (geomagnetic electrokinetograph). This consists of two spaced-out electrodes towed behind a ship. This device registers electric signals proportional to current normal to the ship's directional course. A change of direction will give another set of readings and identify the behaviour of the current. Another ingenious method is just to measure the fluctuations of electrical current produced in submarine telegraph cables by the movement of the water currents. By these means a great deal of information has been extracted about the different submarine currents at various depths, with evidence of what might be regarded as streams and rivers flowing in many directions.

Another type of current is the 'turbidity current'. This is a kind of liquid avalanche. It occurs on the slopes, from the submarine cliffs to the continental shelf. Loose mixtures of sediments and water collect, and periodically an earthquake will send this unstable mixture tumbling down the slopes, starting a mud avalanche travelling like a river in spate as fast as sixty miles an hour. The force of this current hews deep canyons out of the sediment of the continental slag heap.

In addition to horizontal currents, there are vertical currents which bring the cold waters of the bottom layers up to the surface. This high-rise escalator is profoundly important to ocean fisheries because it transports the nutrients from the Earth's greatest compost heap, the sea bottom, to the upper levels in which fishes feed. This accounts for the fact that the richest fishing grounds, or banks, are along the edges of the continental shelf. The mechanism is simple. When the prevailing winds drive the surface currents offshore, the displaced water has to be replaced and, against the submerged cliffs, the colder layers bend upward to the surface. The water comes up from 600 to 1,000 feet below, swirling the nutrients with it. It also considerably modifies the climatic conditions where it happens. The cold upswell chills the atmospheric vapour and produces fogs, like those of the Newfoundland fishing banks or the coast of California.

Weather making

With the knowledge which has been acquired and continuously reinforced, it is obvious that man, always struggling to master his environment, can refine his predictions of the weather by using data provided from the atmosphere and the oceans, but he can also initiate changes.

Much has been publicly discussed about 'artificial weather'. There is nothing wrong with the theory. If a cloud system is building up to the point where it is nearly ready to precipitate, it is possible to trigger it off by artificial methods. To turn into rain, a cloud has to 'nucleate'; it must condense around something, and that 'something' can be a grain of table salt (airborne 'cruets' have sprinkled clouds and caused them to rain) or dry ice (carbon dioxide) or silver iodide crystals, released as 'smoke' from the ground or scattered by aircraft. The clouds will ripen and rain will fall. But in a world already legally harassed by land rights and water rights, litigation over cloud rights is a depressing prospect. It has already happened – farmers have maintained that rainmakers have 'hijacked' clouds en route to their natural fallout. Science fiction (nowadays so quickly overtaken by fact) has already used artificial rainmaking as a form of 'weather blockade': an aggressor intercepts the weather system of the enemy and spills the rain into the oceans, causing disastrous droughts. And, of course, one man's crop-saving rain is another man's ruined holiday. It is now generally agreed that localized rain, over a prescribed area of a few square miles or even a county, will continue to be a chancy business but that regional variations of weather are perfectly feasible.

More urgent is the need to avert the disasters which follow in the tracks of hurricanes. Apart from the loss of human lives (Hurricane 'Flora', 1963, killed almost 7,000 people), the property damage from each hurricane can run into hundreds of millions of pounds. A mature hurricane is a self-sustaining system of atmospheric circulation, a cyclonic 'engine' in which the speed of the wind exceeds seventy-five miles per hour. Most hurricanes have a diameter of between 100 and 800 miles and a lifetime of from one to thirty days. In the Northern Hemisphere, their winds rotate in a counterclockwise direction because of the Earth's rotation. The hurricane derives its energy from the evaporation of warm waters from the surface of tropical seas. The heat generated by evaporation is stored in the form of water vapour, most of which rises in a chimney of huge cumulo-nimbus clouds surrounding the 'eye' of the hurricane. As these clouds grow upward, the water vapour condenses into rain and releases nearly 90 per cent of the latent heat energy. The rest is retained unless the moisture finds an environment in which it can freeze. A lesser amount of energy is generated and later released by a process of spiral rain bands which give the hurricane its characteristic appearance when picked up by radar. In a single day a medium-sized hurricane releases as much energy through condensation as the simultaneous explosion of 400 twenty-megaton hydrogen bombs. About 3 per cent of this energy (equal to twelve hydrogen bombs) is converted into the wind force. Obviously, against such formidable force, there is no way of frontally attacking a hurricane. The only way is to use scientific guerrilla tactics, and, with an inferior force, attack the weak points. In Nature their behaviour was erratic. A hurricane could be observed to develop, collapse, or even reverse its course in half a day. This revealed an internal instability, a susceptibility to 'triggering' influences, like those used in artificial rainmaking.

The object of man's markmanship (like knowing where to shoot an elephant) was the heart of the hurricane, the towering cumulo-nimbus clouds. The hurricane is essentially a huge atmospheric pump which sucks air inward at sea surface and expels it at great heights. Since most of the warm air ascends in the eye-wall clouds, if the wall clouds could be forced apart, the angular momentum of the storm, spinning and moving laterally like a pirouetting ballet dancer across the stage, would be reduced. (The mechanical analogy is the 'governor' which James Watt introduced into his steam engine as the first instance of 'feed-back' automation. His rotating spheres, widening their orbit and operating valves, slowed down excessive speed in his steam engine, and vice versa.) Modification of clouds, as already mentioned, involves converting super-cooled water droplets into snow or ice. Super-cooled water remains liquid at temperatures below the usual freezing point. In the clouds at the top of good-going hurricanes, where the temperature is generally colder than −30° C., water sometimes fails to freeze. At temperatures nearer to 0° C., the water will turn to ice only if there is something in the atmosphere round which it can form.

In the first American attempts at 'stalling' hurricanes, frozen carbon dioxide (dry ice) was used. Dry ice creates ice crystals in a super-cooled cloud by further chilling the adjacent air. One of the first trials was with a moderately strong hurricane, heading from the Florida coast. Its eye was about thirty miles in diameter and its 'chimney' of clouds was 60,000 feet tall. Unfortunately, after 'seeding', the hurricane reversed its course and headed inland to Georgia. There was at the time no way of telling whether this would have happened anyway, but the hurricane chasers were blamed. Until they learned their trade, they were ordered only to tamper with hurricanes which were definitely heading seaward. Moreover, the seeding particles were changed to silver iodide crystals that could be released from a generator which could be dropped like a bomb from an aircraft into the eye of the hurricane. The released crystals were spun by the winds, counter-clockwise round the centre of the storm.

Hurricanes are given coy female names. One such was Hurricane 'Beulah', originating from a hurricane-fecund region east of the Antilles. In August 1963 Beulah was 'bombed'. The operation was successful, but only temporarily. The eye clouds were dispersed but re-formed ten miles away. But the experiments, repeated in successive years with varying and limited success in dispersal, provided measurements (and experience) which explained a lot about the behaviour of hurricanes.

The hundred-billion-dollar question was – If they could not be intercepted and scattered, could they be prevented? In theory, based on such fact-finding, the answer was – Yes, if the evaporation from the sea's surface could be prevented. This would remove the source of heat essential to the creation, and movement, of a hurricane. This, in heat generation, would be the equivalent of pouring oil on troubled waters.

There are ways of preventing evaporation. For example, a body of water (as in the case of reservoirs in hot countries) can be covered by a fine film of cetyl alcohol which prevents convection (i.e. the heat rising from the water) as long as the film remains unbroken. This is all right (though difficult) if the water area is enclosed, but scarcely feasible in the open ocean, with winds and waves. Or the Sun's heat can be reflected before it heats up the water. This could be done by spreading a low cloud of micron-sized particles of aluminium oxide, but its cost would be prohibitively expensive. An alternative with economic possibilities would be organic salts of magnesium extracted from the ocean solution and redeposited on the ocean surface over the hurricane 'breeding' areas.

Short of repressing hurricanes, which is the obvious ideal, the weather satellites orbiting above them have provided a means of watching them being born and tracking their movements. Hurricane prediction, therefore, with warnings for people to protect themselves, if not their fixed possessions, have become a routine. Constant surveillance can be maintained by never-sleeping eyes of wide-angle cameras over tens of thousands of square miles, enabling the forecasters to have a picture of entire hurricane-generating areas every twenty-four hours.

All this is a reminder that we are at the bottom of a sea which is as turbulent and, for those of us who are not sailors, as violent as the aqueous oceans. The 5,000,000,000,000,000 tons of atmosphere are in constant movement, generated by the energy from the Sun and, in turn, generating forces incalculably greater than man has been able to produce even in those delusions of grandeur, the hydrogen bombs.

The greenhouse effect

Although the bomb explosions may not have increased the notorious eccentricity of our weather, as millions of people believe, man's activities are definitely affecting our climate in important respects – and, not in the great sweep of climatic epochs like the Ice Age and its retreat, but in our own time and increasingly in that of our immediate posterity.

Millions of years ago the Sun encouraged the growth of the primeval forests which became the coal, and the organic growth in the seas, which became the oil. The hydrocarbons, converted in the geological pressure cooker, were locked away like the gold in Fort Knox – reserves (in this case, of carbon) withdrawn from circulation. In the industrial era, those reserves have been drawn upon in increasing amounts, to be released into the atmosphere by chimney stacks and exhaust pipes of modern engineering. During the past century the processes of industrialization by the combustion of fossil fuels, have released more than 360,000,000,000 tons of carbon dioxide into the atmosphere in excess of the natural amounts. This continues at a rate of 6,000,000,000 tons-plus per annum. The concentration of carbon dioxide in the air we breathe has increased by approximately 13 per cent above the equilibrium of a century ago, and if all the known reserves of coal, oil and natural gas were burned, the concentration would be ten times greater; in other words, the amount of carbon dioxide would have been doubled.

This is something more than a public health problem (however serious 'smog' may be), and something more than the question of what we breathe into our lungs. The excess of carbon can considerably disturb the heat balance of the Earth by what is known as 'the greenhouse effect'.

The analogy with the greenhouse derives from the fact that carbon dioxide is a transparent diffusion in the atmosphere; it admits the heat of the Sun but keeps the energy absorbed by the Earth's surface and, therefore, considerably modifies the climate. It has been estimated that at the present rate of increase of carbon dioxide (and the rate will increase as more and more nations become industrialized and prosperous enough to multiply automobiles by the multi-million), the mean average temperature all over the world will increase 3·6° F. (2·0° C.) by A.D. 2000. Notice that this is *mean* temperature, averaging out the temperatures, including the extremes of the poles and the Equator. This is a very big alteration in the heat balance. The effects are already apparent not only in the Northern Hemisphere where the industrial excesses are already widespread but also in the Southern Hemisphere. The North Polar ice sheet is already thinning. The winter temperatures of the polar surface waters average about −1·8°, in summer about −1·5° C. Within that range of 0·3° (as compared with the 2·0° C. mean temperature increase which is prognosticated) ice floes lose an upper third of their mass in the summer and acquire a like amount from below in the winter. This equal reconstitution could be affected in two ways – by an increase in atmospheric temperature during the summer and by warming up – a rising into the surface layers of warmer currents generated in the Pacific.

The oceans are an important factor in the carbon-dioxide balance. The oceans contain about 130 million million tons of carbon dioxide – about fifty times as much as the air. Some of the gas is dissolved in the water, but most of it is locked up in the carbonate compounds. The oceans exchange about 200,000,000,000 tons of carbon dioxide with the atmosphere each year. When the atmospheric concentrations rise, the oceans tend to absorb some of the surplus, and when they fall, the oceanic reservoir tends to replenish the air.

Both the atmosphere and the oceans continuously exchange carbon dioxide with rocks and with living organisms. They gain carbon dioxide from the volcanic activity and from the respiration and decay of organisms; they lose carbon dioxide to the weathering of rock and the photosynthesis of plants. The organic life of the sea, using the light rays of the Sun, converts carbon dioxide to its processes of growth, but plants also exhale carbon dioxide.

As these processes change pace, the carbon dioxide in the atmosphere also changes, altering the radiation balance and raising or lowering the Earth's temperature.

The carbon-dioxide mechanism could itself account for the periods of glaciation, when, for instance, volcanic activity had halted or when an increase in vegetation, in swamp conditions, was locking up the carbon dioxide which would eventually become the coal deposits. In a relatively short time (geologically speaking) the carbon dioxide in the atmosphere could be reduced by 50 per cent, which represents a drop of 6·9° F. (3·83° C.). This would cause glaciers to spread across the land surface, locking up moisture in the ice and thus lowering the oceans (by 5 to 10 per cent at the height of glaciation). The shrunken oceans would accumulate carbon dioxide in excess, which they would return to the atmosphere, causing it to warm up again and so start another cycle. But we are talking about cycles of the order of 50,000 years. Consider what could happen in fifty years. It is known that plants borrow 60,000,000,000 tons of carbon dioxide for photosynthesis.[1] This loan is almost completely repaid annually by respiration and decay. The Earth's hot springs and volcanoes spill out about 100,000,000 tons of carbon dioxide, but weathering rocks recapture that in forming carbonates. Modern man, however, is shortening aeons into decades. In addition to the 6,000,000,000 tons of combusted fossil fuels, which is an entirely unnatural increment, man's agricultural activities are releasing 2,000,000,000 tons. Grain fields and pastures store much smaller quantities of carbon dioxide than do the forests that they replace, and the cultivation of the soil allows large quantities of carbon

1. Gilbert N. Plass, 'Carbon Dioxide and Climate', *Scientific American*, July 1959, pp. 41–7.

dioxide produced by bacteria in the soil to escape into the atmosphere. Not all this industrial–agricultural carbon dioxide remains in the atmosphere. Some is recovered by plants, but the vegetation cycle of absorption and respiration does not take care of the fossil fuel excess. Most of this, if it were halted, would be absorbed by the ocean waters themselves. Studies have shown that the volume of carbon dioxide in the oceans comes into equilibrium with the carbon-dioxide pressure of the atmosphere in about 1,000 years.

But it is only necessary to look at the present reliable records of temperature and fossil-fuel consumption and project them into the future to predict the changes in climate. There is this 13 per cent carbon increase during the past century. The carbon-dioxide theory correlates this with a rise in the average temperature by 1° F. (0·55° C.). This is precisely the recorded average increase all over the world in the past century. But, if we take the records of more recent periods, for England, Scotland, Austria, Iceland, Sweden, and Norway, we find that the mean temperature for the latitudes has risen over a period of fifty years (of increasing combustion) by 1½ to 2° F. (Like most scientific theories, this one was under attack almost as soon as it became accepted; in 1965–6 a committee of the U.S. National Academy of Sciences suggested that since most of the increased carbon dioxide would be absorbed by the oceans the effect on temperature would be quite small.)

Already there is evidence of substantial fish migration following the warming up of the seas and the northward increase of the growth of phytoplankton and zooplankton on which they feed. On land, the snow line is retreating northward. In Scandinavia, land which was perennially under snow and ice is thawing, and the arrowheads of over 1,000 years ago when (in a natural cycle) the black soils were last exposed have been found.

This black soil is another factor in the heating up. It is manifested in the Sub-Arctic of the Western Hemisphere where the snow line is retreating appreciably northward. When land is covered with snow and ice, the white surface reflects the Sun's rays and reduces the warming up so that the permafrost – frozen rocks and subsoil – remains. If, however, there is an upward variation of surface temperature and the thaw continues a little longer each summer, the exposed black soil, being a heat absorber, holds the heat longer and it penetrates deeper and affects the permafrost so that the bottom falls out of the frozen muskeg and drainage takes place – again affecting the climate. It is a double action – atmospheric temperature increase and black absorption retaining heat – so that the process accelerates. (The Chinese in recent years have speeded up the melting of glaciers by scattering carbon black on the snowfields, in order to get more water.)

Glaciers are retreating either by melting or by breaking up, at coastal level, into icebergs. This, as has been pointed out, adds additional water to the ocean levels. It will, however, have other effects as well. Rivers which originate in glaciers or permanent snowfields will increase their flow. Vast amounts of water are at present retained (in the Himalayas, for example) by ice barriers, the melting of which would be like the breaking of massive dams, with incalculable results in lower-lying regions. The Swiss avalanche disaster of 1965 followed the collapse of an ice-retaining wall.

It is scarcely necessary to remind people of what a few feet rise in the sea level (let alone the ultimate 200 feet which would follow the complete liquidation of the land-held icecaps) would do to our geography. Suffice it to say that it would be unwise to take a ninety-nine year lease on coastal flats!

Apart from interfering with geography, man's processes are going to affect climatic distribution. These changes have to be understood and foreseen – and that can only be done by a combination of sciences, not the isolated study of component parts. The clearing of forests which absorb and transpire moisture changes local climates. Moreover, when the destruction of the vegetation cover removes the umbrella of foliage and the sponge of the undergrowth which seeps the water into the ground-water system, the scouring rains flush off the soil in flash floods, to raise the silt beds of the rivers, and the bared hillsides shed the water as from a tiled roof. This runoff into inadequate 'gutters' causes 'disaster areas' and, at the same time, city droughts, because of the failure of the water to seep underground and replenish the wells supplying reservoirs that serve the cities.

Some authorities regard interference with the upper atmosphere as more serious than the aggregation of carbon dioxide, because the effects on the rain systems may be more immediate. This interference includes high-level nuclear explosions.

As was indicated earlier, the physical 'thrust' effect of the most powerful hydrogen bomb is fractional compared with a modest hurricane, but the fallout, about which so much was heard during testings, is another matter. We have seen how 'weather' builds up around particulates – like the super-cooled clouds 'latching on' to crystals of dry ice or silver iodide or table salt. The particulates of the bomb explosions provided such nuclei and the radiostrontium came back to Earth in rain. But there are other man-made disturbances of the natural equilibrium: absorbing layers created by rocket exhausts or the deliberate introduction of copper filings or the artificial disturbance of natural layers (like the 'Rainbow Bomb' in the Van Allen Belt). The cloudlike condensation trails produced by high-flying aircraft may exert a more immediate effect on the climate than carbon dioxide. The cloud cover is the most important component in determining the Earth's reflectivity and it serves as a natural thermostat to keep the world's temperature within tolerable limits. Whenever cloud cover increases, less solar radiation reaches the Earth and vice versa. The average cloud cover over the Earth is estimated at close to 55 per cent; the average reflectivity and cloud cover together is estimated at 35 per cent. A 1 per cent increase in cloud cover causes a 0·4 point increase in reflectivity which by reflecting solar radiation causes a decrease in temperature of 0·7° F. Increasing condensation trails as the high-speed, high-flying aircraft multiply will alter the proportions.

One of the functions of weather satellites is to make measurements of the energy which reaches the Earth and to note the circumstances in which it is lost. If records show over the years that this debit or credit is being affected, then counter measures will have to be taken. The energy budget is more important than ledger-balancing operations of ministers of finance. The lives of people and the fates of nations will depend on patterns of rainfall. Man's present activities can by inadvertence (or ignorance) change the rainfall distribution. This will increase precipitation in some areas now arid, and cause the drying out of areas now fertile.[2]

The curse of the albatross

'Water, water, everywhere, nor any drop to drink', lamented Coleridge's Ancient Mariner. The curse of the albatross which fell upon him can fall on whole communities.

As the world's population multiplies and urbanized society sprawls upward and outward in water-consuming, water-shedding, and water-polluting cities, the science of hydrology becomes vital in a literal sense, since it is a matter of life and death. Other sciences and the technologies derived from them can be invoked (like using nuclear-energy power stations to desalinate sea water), but this is a penitential way of repairing the neglect of proper study and proper handling of a natural commodity which is more precious even than food, since food itself depends upon it.[3] Moreover, it takes a person at least a fortnight to die of hunger but only three days to die of thirst.

The hydrological cycle on which we depend for our survival is fairly well known. Basically it operates by evaporation and precipitation. The heat of the Sun causes bodies of water, such as oceans, to vaporize. The convection currents carry the steam upward until cooling, at higher altitudes, causes clouds to form. The clouds are decanted as rain which falls back to Earth to water the soil; to sink into the ground to replenish wells and spring sources; or to be stored in aquifers, water-holding layers below ground. Trees and vegetation extract some of this moisture and transpire it again into the atmosphere. The land waters form streams and rivers which flow into lakes that are reservoirs for other rivers which flow eventually to the oceans. If, like the Dead Sea, they have no outlets to the oceans, they themselves become saline through the accumulation of salts left by constant evaporation. And so the cycle continues. But that does not mean

2. Cf. Morris Neiburger and Harry Wexler, 'Weather Satellites', *Scientific American*, July 1961.

3. Cf. Ritchie Calder, *World of Opportunity* United Nations Publications, 1963.

that the pattern of rainfall has always been the same, in terms of geography.

The great arid expanse of the Sahara was not always so. In the 1950s, French scientists investigated a large area of the Tanezrouft, in the heart of the Algerian desert. It was gravel desert which had always been regarded as the most hopeless kind. Under the gravel they found soil which had been protected by the stony cover from being blown away by the winds and from being burned out by the sun. It contained the fossil remains, and pollens, of vegetation similar to that of the contemporary Mediterranean littoral, evidence that seven thousand years ago there was a lush growth covering, in this particular region, an area of 50,000,000 acres, which is as big as Great Britain. There is other substantial evidence to indicate that at that period the source of the Niger was in the Atlas Mountains and that the main river, as we see it now, was only a tropical tributary. The intervening region dried out and the rainfall became minimal. The rivers which flow from the Atlas became wadis, spilling in brief violent spates to spread over the thirsty desert and evaporate, or, as is now known from the studies of the Albienne Nappe – the vast underground aquifer under the Sahara – to seep down into the intercalary continental nodular sandstone layer.

The studies of underground water systems, such as the Albienne Nappe, are quite recent. The existence of the 'nappe' was postulated early in the twentieth century but was treated as a geological myth. When French hydrologists insisted it existed, they were derided – as the oil men were when they first started to prospect for oil under the 'sand sea'. They persisted and in 1949 a drilling at Zelfana in the northern Sahara produced a water 'gusher' which, under its own pressure, sprayed upward 300 feet. The abundant proof, however, came with the drilling for oil when, at Hassi-Messaoud and elsewhere, the drillers going down 11,000 feet for oil struck water at 3,000 feet. The oil-boom towns were thus able to have artificial swimming pools, district-cooling in the form of a fine spray, and fresh food, since the artesian waters could be used to extend the irrigation systems of old oases and to create man-made ones. The Albienne Nappe, with well-identified sources of replenishment where the intake can be measured, forms a basin under the Sahara, from the Atlas Mountains far to the north and from the Mauritanian, Atlantic, coast. It extends under Tunisia and Libya. In the east there is a similar water-bearing formation, the Nubian Sandstone Layer, which, fed from equatorial Africa, extends to the Mediterranean, with a wind-scoured outcrop in the Quattarah Depression, half the size of Wales. The Nubian Layer is the source of the water which has perennially supplied the great oases which lie in the desert depressions, and it provides an underground tributary which flows under and seeps up into the Nile. The aquifer structures extend under the Red Sea and reach the Persian Gulf. Similar underground reservoirs have been found in the Indian subcontinent and under the Gobi Desert, long regarded as the most hopeless of all the empty places.

Mining for water

It became apparent from such findings that the water in the lakes and rivers on which mankind has mainly relied for its supplies is in fact an insignificant fraction of the world's H_2O. Obviously the oceans and their brine account for most of it – 97·2 per cent or 317,000,000 cubic miles. Ice caps and glaciers have locked up 2·15 per cent or 7,000,000 cubic miles. The total liquid water in land areas is 0·635 per cent or 2,070,000 cubic miles. Freshwater lakes account for only 30,000 cubic miles; saline lakes and inland seas, for 25,000 cubic miles; rivers and streams, for a mere 300 cubic miles; soil moisture and near-surface groundwater, for 16,000 cubic miles; groundwater within a depth of half a mile, for 1,000,000 cubic miles, and deep-lying groundwater, for another 1,000,000 cubic miles. Thus the amount of water in subterranean storage is 6,000 times greater than the amount of water in the rivers at any given moment of time.

Subterranean storage has the great advantage that, unlike surface reservoirs, there is no loss by evaporation and, if it can be tapped by tube wells, the water supply can be used where and when required. Apart from natural deposits of water, it is possible to use porous rocks as artificial underground reservoirs by pumping surface waters (during the wet season when they would otherwise flush to the sea) into well formations. It is now a relatively simple matter to study the holding capacity of such formations by the use of radioactive tracers which can be harmlessly introduced into the input waters and any seepage can be instrumentally detected. The storage space available below ground, in porous materials, is nearly everywhere in the world very much larger than even the largest man-made surface reservoirs. One of the objects of the International Hydrological Decade which began in 1965 was to find out a great deal more about these natural storage possibilities.

Some of those deposits are geologically sealed off in 'water bottles' which will never be replenished, like oil wells which empty. Others fill and empty very slowly (the subterranean tributary of the Nile progresses thirty feet a year). Others are rivers which flow into the ocean below sea level, like the one which debouches from the limestone formations of Lebanon into the Mediterranean.

This kind of knowledge has not been acquired by dowsing, however effective the twisting ash twig may be; it is the result of hydrological research, carefully conducted and internationally shared. It really began to mean something as a result of the Arid Zone Programme of Unesco, which brought together and encouraged experts from all the desert regions of the world. The water aspects of that programme were consolidated and extended by the International Hydrological Decade.

The water balance sheet

If human enterprise is to go on interfering with the cycle and to draw upon the underground reserves, especially in places where the local precipitation is minimal, a careful ledger will have to be kept. Extracting underground water means drawing on water capital in the absence of water income. The progressive lowering of the water levels in the aquifers may have consequences more serious than the subsidences which follow mining of ores. It is, however, possible to learn the nature of the underground sources by assessing the amount of tritium in any water discovered. Tritium is 'triple hydrogen', an isotope which exists in Nature and which is deposited by atmospheric precipitation. By metering the amounts of tritium (and knowing its half-life) it is possible to determine whether the supply is being replenished, at what intervals, or not at all. For instance, water sealed off in geological times would have no active tritium. Another way is to trace the amounts likely to be going into the aquifer. For example, the nodular sandstone formation of the Albienne Nappe forms a vertical outcrop in the basin between the High Atlas and Saharan Atlas Mountains. It is like a funnel, leading down into the horizontal formation under the desert. The annual intake from this surface source can be fairly accurately measured, and if users were to confine their demands within this known 'income', there would be no problem and the use could be increased as new sources of 'income' were discovered. Nothing in our previous history of prodigal use of water and the reckless squandering of supplies offers much reassurance. Moreover, there will be the excuse of imperative requirements as the population increases.

The rapid increase in population, doubling within thirty years, with the continued development of agriculture in lands previously uncultivated, and the industrialization of countries hitherto undeveloped, means a radical transformation in the pattern of water usage.

The amount of water used *per capita* varies enormously from country to country – from ten gallons a day in some under-developed countries to 1,800 gallons a day in the United States. Of the latter figure about 6 per cent is for domestic uses, the 94 per cent being shared by agriculture and industry. It takes 300 gallons of water a day to produce the grain needed for a two-and-one-half-pound loaf of bread and forty tons of water to make one ton of steel. Some 60 per cent of all water used in irrigation evaporates from the canals or from the soil or is transpired through the leaves of the plants. City waste and industrial cooling represent large amounts of water, but this water finds its way back (however much polluted) into the rivers or into the ground and figures in the water balance sheet as a continuing asset. But with the wholesale contamination of the rivers, such water can only be recovered at enormous expense by chemical treatment. Many great cities are already drinking their own sewage.

The demand for water will continue to increase enormously. Domestic consumption becomes exaggerated by the amenity needs – baths, toilets, street cleaning, car washing – by the need for fire fighting, and so on, but while quantitatively large, it is still a small fraction compared with the water needed for steel-making, papermaking, and chemical, rubber, and petroleum production, which not only use a great deal but contaminate much more.

One way to intensify the food production necessary to maintain the increasing population is by irrigation. This means storage – usually in surface dams; and, as in the case of the Aswan High Dam, in a region of high atmospheric temperature, the loss by evaporation can be as high as a third of the storage capacity, plus the loss through canal distribution. Predictably the demand for water will be four times as high by A.D. 2000 as it was in the 1960s and, while the total stream flow will still be many times larger than the demand, the economic cost of its use will be far higher. Storage, recycling, transportation (with the prospect of pipelines from the Arctic to the United States) and, eventually, large-scale desalination of sea water will make water an expensive commodity. More and more attention will be directed to 'mining' the groundwaters and, as has been pointed out, this will need much more exact scientific knowledge not only to determine the supplies but to conserve them. Hydrology has thus become one of the critically important earth sciences.

11 Lithosphere

A sobering reminder

The International Geophysical Year demonstrated the essential unity of the earth sciences – a unity often disastrously disregarded in practice. One such disaster was writ large in Pakistan.

The President of Pakistan appealed in 1961 to the United States for help in a desolate situation. President John F. Kennedy responded by sending a mission of 20 specialists from as many disciplines, with a rear echelon of computers at Harvard University. That cross section of the sciences was necessary to study the mistakes which had been made because of an anarchy of science in the first instance.

The soils of the Punjab and the Sind were created by the Indus and its tributaries, the Jhelum, the Chenab, the Ravi, the Beas, and the Sutlej. In the alluvial soils they had laid down, one of the earliest civilizations flourished, the relics of which are still to be found at Harappa and Mohenjo-Daro, settlements which existed 5,000 years ago.

In the nineteenth century, the British began a big programme of farm settlements in lands which were measurably fertile, but with low rainfall. Dams and distribution canals were constructed, with the pinchpenny proviso that the master canals were unlined – after all there was plenty of water in the Indus. What matter if 40 per cent of the distribution water leaked underground? Those engineering works were massively expanded and the irrigation widely extended after Pakistan became independent. The 23 million acres watered by canals became, in 1960, the largest irrigated region in the world. West Pakistan was essentially a productive region won by man from the desert. The lives as well as the livelihoods of 30,000,000 people depended upon it. Over 40,000 miles of canals were dug into the surface of the Indus Plain. What had been 'economical' practice was continued – the canals were unlined. Apart from the water which was spread over the fields, the 40 per cent seepage was exaggerated on a greater geographic scale. The Indus incline is 700 feet in 700 miles – one foot per mile. The principle was 'inland delta' drainage; the spread water would find its way back into the river and be carried to the Arabian Sea. This drainage did not happen. The result was that the water table rose.

Before the introduction of canals the water table was well below the surface. Only in some rare cases in areas close to the river was it between five to fifteen feet. After seventy years of irrigation the picture changed completely. The water table over vast areas rose close to the surface. Low-lying land became waterlogged, drowning the crops. In other areas the water seeped upward from the water table to the surface, where it evaporated, leaving its salts to accumulate in the upper layers, poisoning the crops. At the same time, the irrigation regimen, which used an average of one and one half feet of water per year, spread out a layer of surface water, with its own dissolved salts, so thinly that by evaporation it left its own surface crust of salt.

This combination of waterlogging and salination produced a deterioration at the rate of 100,000 acres a year. (In one district in Punjab, in the upper reaches of the Indus, the extent was more than 50 per cent of the culturable land.) When the President of Pakistan made his appeal in 1961, one acre of land was going out of cultivation every five minutes; local increase of population required that ten more mouths had to be fed in that same five minutes.

The investigators from the manifold disciplines found what the diversified experts had ignored. No one could criticize the engineers who had done a magnificent professional job in constructing the dams and in distributing the water. (It was the fault of the policy makers, not the scientists or technologists, that the master canals were unlined.) The soil physicists, the soil chemists, and the agronomists, each in their own efficient way had satisfied themselves that the alluvial soils (carefully analysed, horizon by horizon, in inspection pits dug in the desert) could support rich crops. And they did. But somehow the hydrology had been misread or not read at all. And, one might add, archaeology had been neglected too. Mohenjo-Daro had flourished and had declined, with a message in its ruins of pertinence to contemporary developers of deserts. Moreover, in the Mesopotamian lands, between the Tigris and the Euphrates, and in the Aral Sea civilization of the Oxus, farming had contrived to overcome waterlogging and salination. In ancient Sumer and Babylon, with no mechanical drainage and relying on 'inland-delta' drainage, agriculture was maintained, according to temple records, for 700 years. They used 'plant drainage'. They alternated their agricultural crops with deep-rooting, thirsty, salt-loving weeds. These dried out a barrier layer in the soil between the water table and the surface, to prevent upward seepage. Therefore the farmers were contending only with surface evaporation salts and using enough irrigation water to leach those salts and sluice them underground, below the root levels. One must remember, however, that archaeology, with all its abundant lessons of trial and error, through centuries and millennia of our ancestral experience, is not really regarded as a science; it is a limbo discipline, regarded as descriptive and not on a par with the exact sciences.

The upshot of the concerted scientific investigation was, one hopes, salutary; it was certainly sombre. To repair the damage done by diversified sciences, measures were necessary which would cost $2,000,000,000 and take twenty-five years. It was recommended that the restorative action should include not only the reclamation of the salted waterlogged lands, but agricultural and social improvement as well. For example, to redress the water balance would require tube wells. These could be used to supply irrigation water by drawing from the water table, thus lowering it in the process, provided there was a limit, on a proper budgeting, to the influx of irrigation water from the Indus itself. To provide the power for the pumping, electricity would have to be produced, hydroelectrically and by thermal stations using the natural gas discovered in the region. This electricity could also produce fertilizers to enhance the condition of the soil and bring it back to health and increased productivity. This productivity would require improved and adapted crops, with proper training in how to use them – seed stations, educational courses, and agricultural extension services. With all this there would have to be developed a new social pattern of villages and local industries. In other words, the lesson of Pakistan was not only the need for the coordination of the natural sciences and their related technologies but the bringing in of the social sciences as well.

Inventory of natural resources

The earth sciences have the search warrant to seek out the knowledge of the world's natural resources, as well as to study the forces which influence the planet and the lives of the people who live on it. On the inventory of those resources and the uses to which the decision-makers put them depends the progress of mankind.

Over 1,200 minerals of the Earth's crust have been inventoried and the list increases. Most of them are mineralogical curiosities, but further inquisitiveness on the part of the scientists and further inventiveness on the part of the technologists can make many of them important. For instance, the ancient Egyptians used malachite from the Sinai as a cosmetic (for eye shadow), but over 5,000 years ago some experimenting metallurgist discovered that malachite could be smelted into a particularly hard copper. The Roman Emperor Nero used a lens of crude emerald (beryl) as a quizzing glass with which to watch the gladiators in the arena, but modern metallurgists convert it into the metal, beryllium, invaluable in atomic reactors.

Nuclear developments, apart from giving a new importance to uranium and thorium as fuels, have given a new significance to minerals which were but poorly regarded. Lithium, for example, has come into its own with the discovery of thermonuclear energy. As lithium-6-deuteride, it provides a source of deuterons, the double atoms of hydrogen, which can be fused to form atoms of helium, with a surplus of energy a thousand times greater than that released in the splitting of atoms. Cadmium, boron, and hafnium, which metallurgists had previously found useful only in certain alloys, became self-important. They absorb neutrons and can therefore be used as the 'brakes' of a reactor. When a reactor becomes 'critical', that is to say, when it is sustaining a continuous chain reaction, turning U^{235} into plutonium, there is a risk that the flow of neutrons may become excessive. Rods of cadmium, boron, or hafnium can be automatically dropped into the reactor to 'swallow' the surplus neutrons. Zirconium is an analogue of hafnium, but has the opposite property, in that it does not readily absorb neutrons and therefore forms a useful alloy with plutonium.

The existence of minerals and knowledge of their properties do not necessarily make them available. They may not be accessible in quantities which make them economic, or they may be unavailable because the mining, separation, and refinement make them too expensive, or because, at any given moment, the means of their conversion may not exist commercially. Aluminium, that now-universal metal, at first had no economic value. It became a usable metal only when electricity became abundantly available for the electric reduction of bauxite. Today minerals which were once intractable can be reduced and refined by vacuum distillation and melting, by high-frequency electric furnaces, or by use of intense solar energy focused by mirrors. Where metals would not respond to smelting, they can now be processed by powder metallurgy, in which the particles of the elements are compressed to form a solid, or by sintering, in which the powdered chemicals are compressed and heated, forming densely packed briquettes. The technological heirs of those dissimilar craftsmen, the potter and the smith, have come together with the merging of ceramics and metal techniques. Uranium dioxide is a ceramic, but by combining it with plutonium dioxide the resulting material becomes a useful nuclear fuel unit. Similarly, uranium carbide plus plutonium carbide can serve as a ceramic. Conversely, there are 'cermets' (ceramic-metals) in which the oxides and carbides are distributed in discrete particles in a metal. Metallic crystals, otherwise brittle and unstable, become practical when embedded in plastics.

Silica, as sand and sandstone, is one of the most widely distributed elements on Earth. In remote times it was found that it could be fused into glass, but its present range of usefulness has been vastly extended by the development of silicones, used in products ranging from shoe polish to the giant tyres of jet airliners, and by the production of elemental silicon from silica sand. This provides the semiconductor of electricity, now used in transistors and in solar batteries which convert the energy of the Sun into electricity and have been used to provide power for the transmission of radio signals from satellites and space probes.

The semiconductors exploit the physical fact that electrons (and therefore electric currents) will flow between certain incompatible crystalline solids, e.g. between a flake of germanium crystal and a fleck of indium. These in combination, although no bigger than a match head, can act as a valve as efficient as a glass vacuum tube. Silicon or selenium can serve as well as germanium.

New metals

With new requirements, 'new' metals have to be considered. Supersonic speeds for aircraft and the re-entry of space vehicles into the Earth's atmosphere, the friction of which generates the heat which combusts meteors and makes them shooting stars, impose new conditions. Titanium, a laboratory curiosity a few years ago, becomes a metal prescribed for the conditions created by supersonic speeds at which steel does not provide the high-tensile material necessary. For answers to special requirements, the metallurgists have turned to refractory metals such as zirconium, niobium, molybdenum, or tungsten. Metals are being used as composites which give the ductility necessary for forming them into required shapes, less brittle and retaining their shape and strength at high temperatures. These composites are not conventional alloys; they are globules of metals, with low melting points, imbedded in refractory metals.

These are the refinements of increasingly sophisticated advanced technologies. But the demand for 'old' metals continues to increase. Dr Harrison Brown, of the California Institute of Technology, at the United Nations Conference on the Application of Science and Technology for the Benefit of the Less Developed Areas (U.N.S.C.A.T., Geneva 1963), pointed out that with the increase of population and industrialization the quantity of steel in use in the world will grow in the predictable future to 70,000,000,000 tons (compared with the 2,000,000,000 tons in use in 1963 in the U.S., the world's biggest producer). Associated with the steel in use, in the form of machinery and structures of various types, there would be vast quantities of 'conventional' metals. It can be reckoned that for every ton of steel about 40 pounds of copper, and similar quantities of other metals such as lead and zinc, will be in use. Thus the demand for copper, lead, and zinc will be of the order of 1,000,000,000 tons. The demand for minerals and metal will increase at three times the rate of the population growth because countries which had been held back will be trying to catch up, to match the industrial prosperity of the scientifically and technologically advanced countries. And they have, in terms of untapped resources, the rich ore beds which the advanced countries, depleted by their own rapid advances (and mineral profligacy) will want. Those natural ores are their capital. They can squander it.

> We must recognize [said Harrison Brown at the Conference] that we are approaching the time when men the world over will have to gain their livelihood from the lowest common denominators of the Earth's crust – air, sea water, ordinary rock, and sunlight. Air will provide us, on an expanding scale, with nitrogen for agriculture. From sea water we will obtain fresh water, metals, and salts. Ordinary rock will provide us with the majority of metals, phosphorus, and carbon, and the greater part of our energy. Sunlight will continue to provide energy for agriculture and will provide space heating together with power for specialized purposes.[1]

Laboratory without walls

Not so long ago one might have said that the 'earth sciences' were bounded by the gravitational fence, that everything within that imprisoning force, which *kept* man earthbound, could legitimately be labelled 'geo-' and everything beyond could be prefixed 'astro-' or 'cosmo-'. That is scarcely true today when there is a continual interrelationship between the earth and space sciences. Indeed, as has been pointed out, the Space Age was initiated by the satellite programme promoted by International

1. Quoted in Ritchie Calder, *World of Opportunity*, U. N. Publications, 1963, p. 19.

Geophysical Year because it was deemed necessary to 'step out and see' what the Earth and its envelope looked like from outside and to study the physical forces which impinged on 'Planet Number Three'. We cannot set the offshore limits of our globe by reference to the magnetic belts because our geophysical conditions are influenced by the Sun and by particles originating in outer space. By extension in the other direction, the earth sciences are involved not only with the core of the Earth but with the core of the atom.

The study space of the earth sciences, therefore, is a laboratory without walls in which we may set up movable partitions. As a matter of scientific housekeeping, we enclose the various physical and natural sciences in closets, but as interdisciplinary neighbours. Intercommunicating with all of them is mathematics which, as the word 'geometry' reminds us, began as earth measurement.

Historically, the earth sciences were observational, with measurements based on empirical techniques. Over thousands of years data were collected from which, by very slow process, laws and principles were formulated. Only in the twentieth century did the technical resources come to be applied to those observations and to the refinement of measurements, completely transforming their value through simultaneous increase in their number and accuracy.

The possibility of conducting a great number of internationally coordinated observations all over the planet, of collating the results rapidly, and of processing them by almost instantaneous statistical and mathematical procedures meant that a completely global picture could be obtained of natural phenomena, which disparately could not be fully comprehended. Nor was it enough to set up networks of observation posts on land or afloat; stations at increasing distances from the Earth were needed – balloons, rockets, artificial satellites, and space probes; so were stations below the surface of the land or of the sea.

Let us consider the effects on the old classifications. Geodesy is the oldest branch of geophysics; its purpose was to determine the exact shape of the Earth and its field of gravity. These two questions are inseparably linked since the Earth is a spinning mass. It is now known that this mass is far from being homogeneous and its shape not merely the 'flattened orange' of our schoolbooks. By study of the orbits of artificial satellites and by using satellites in orbit as triangulation points, and with time measurements of an accuracy of one-thousand-millionth part of a second, an exact picture of the globe was obtained. With increasing precision of measurements, it became possible to obtain data on 'land tides', i.e. the gravitational pull on the continental masses, and to learn about the elasticity of the Earth's crust.

While space exploration was revealing more and more about the Earth's general magnetic field, new precision instruments could detect anomalies in the magnetic field at the Earth's surface and at the sea bottom. This was of more than academic importance because the knowledge lent itself to mineral prospecting at hidden depths within the Earth's crust. Coupled with that was the study of rock magnetism, with its bearing upon the Theory of Continental Drift.

Other geodetic studies have concerned the radioactivity of the rocks and the temperature distribution in the crust. The magnitude and direction of the flow of thermal energy through the Earth's crust can be evaluated and compared with the energy coming from solar radiation.

Geology acquired a new status. The traditional picture of the geologist (as a person in hobnailed boots, chipping rocks with a hammer on the outcrop formations, or gratefully following the construction workers into railway cuttings or tunnels, or hunting for fossils in coal mines) was changed. He could recruit the help of airmen to map the terrain by aerial photography and he could, stereoscopically, recognize telltale formations which no amount of foot slogging would have revealed. He could use airborne scintillometers (to detect radioactive rocks), gravimeters, and magnetometers. He could use man-made explosions to produce artificial earthquakes to give him seismic identification of rock stratification. Instead of being dependent on the indulgence of commercial prospectors, who might or might not give him the cores of their drillings, he could claim academic (non-profit-seeking) rights to his own borings so that he could begin to build up new geological maps. (When the United Nations Economic Commission for Asia and the Far East tried, in the 1950s, to compile a geological survey of south-east Asia, it was found that 95 per cent of the area had not been geologically surveyed on a scale which could even provide a guess at the hidden resources; 5 per cent had been so surveyed and of that only half had been exploited. That 2½ per cent had been the fabulous wealth of the Indies!)

Whereas in the past the study of rocks had been largely a matter of 'dig and try', the demand of advancing technology for more specific minerals meant that the science of the actual structure of the rocks, i.e. petrography, assumed a new importance. From an observational discipline it became an experimental one. With advanced experimental techniques and laboratory technology, e.g. high-pressure chemistry, it became possible to produce the true synthesis of certain minerals – as Nature had once compounded the elements before they were regurgitated in conglomerate ores.

The oceanographers had a lot to teach the geologists, by studying the present-day ocean beds and showing how sedimentary rocks were built up and, in turn, the geologists had a lot to teach the pedologists, who study soils.

The hierarchy of science

Nothing in the earth sciences happens in isolation. If we are to control our physical environment which dictates our living environment, it can only be done on an interdisciplinary basis. Observation has to be checked by experiment. Experiment heightens observation. Observation and experiment increase information. Information suggests new applications. Applications accelerate technology. Technology reciprocates by providing new means of observation and experiment.

Customarily, science, or the scientific hierarchy, is divided into four categories:

Pure, or academic, research is the pursuit of knowledge for its own sake. It is mainly the work of an individual, or the group he leads. There is no 'payoff' except in the emoluments of distinction, or, possibly, the Nobel Prize. The 'pure' scientist has to justify himself only before a jury of his own peers. He is judged not by the usefulness but by the integrity of his work. He is the Maker Possible.

Oriented fundamental research is still basic science, that is to say, the scientist is still directly questioning Nature, seeking to extend knowledge and understanding, but he is not a free agent, indulging his private curiosity. He is constrained within a frame of reference. He is, for example, studying chemistry at high pressures, without assuming that he is going to discover polythene; or studying gases at high temperatures, without envisaging the jet engine or the rocket in which those gases will operate. He is finding out information which will be important in a general field and likely to have some foreseen application. In big corporations this is called 'speculative research'. It might be called 'altruism with a motive'. The scientist is likely to have adequate research facilities and endowments, or contracts. He is the Maker Probable.

Applied research is programmed research. The target is specified and results are expected. The predicted yield is the measure of the support. The scientist is held accountable in the annual report. He is the Maker to Happen.

Development is really technology, but 'development' somehow keeps it in the scientific hierarchy and away from the 'rude mechanicals'. This is the transfer of laboratory results, through the pilot plant, to the shop floor. The 'R. & D.' (linking 'Development' with 'Research' keeps it respectable) scientist is the Maker to Work.

In the earth sciences, the laboratory without walls, are to be found all those categories: from those who ask the questions about the nature of our planet to those who extract the minerals and oils and convert them not only into the fuels of transportation but into 50 per cent of all the chemicals on the world's market.

12 Microcosmos

The invisible universe

No one has ever seen an atom. Yet a nuclear physicist conversationally will reconstruct for you the invisible universe with the apparent confidence of an astronomer building a planetarium. The operative word is 'apparent', because if there are doubts and debate about the nature of the macrocosmos, there are greater doubts and greater debates about the nature of the microcosmos. Paradoxically, the more information the scientists get from their gargantuan instruments the more questions they raise.

When the ancient cartographers had only vague ideas about some unexplored *terra incognita*, they labelled it, 'Here be dragons'. The nuclear physicists in exploding the atom released the dragons before they had properly explored the territory.

On 16 July 1945, the atom exploded with a cataclysmic force a thousand times more violent than the most powerful chemical explosive then known. This was 'fission', the splitting of heavy atoms. Presently there came 'fusion', the welding of the lightest atoms, with the resulting thermonuclear force of the H-bomb, a million times more powerful than the chemical explosive. When Einstein realized the destructive use to which his equation $E = mc^2$ (energy equals mass multiplied by the square of the speed of light) could be put, he said, 'I wish I had become a blacksmith.'

To understand why, in 1945, the bomb became technologically possible we have to go back even before 1905 when Einstein, unwittingly (in $E = mc^2$), said it could be. Röntgen had discovered X-rays. J. J. Thomson had discovered the electron. Becquerel had prepared salts of an odd element, a laboratory curiosity, a mineral nuisance, uranium, and had discovered radioactivity – rays spontaneously generated by the element itself, not imbibed and re-emitted like the luminescence of fluorescent salts; Marie Curie had isolated radium; Ernest Rutherford had already asked the multimillion-dollar question, 'How can atoms (supposed to be ultimate particles of matter) give off rays?' He had been the Sorcerer's Apprentice to J. J. Thomson (in the researches which had led to the discovery of the electron) and went to Canada at the age of 27, to become Professor of Physics at McGill University, where, according to Sir Arthur Eddington, he initiated the greatest change in our ideas of matter since Democritus, 400 years before Christ.

He established, in the first instance, the existence of two emanations, distinct from the electrons with which he had worked with Thomson. These were the alpha rays and the beta rays (the 'A' and 'B' of the Greek alphabet). Alpha rays we now know to be the nuclei of helium, and beta rays we now know to be streams of electrons released from any source. Working on thorium, he established 'half-life'. (In his own words [1906]: 'In the first 54 seconds, the activity is reduced to half value. In twice that time, i.e. in 108 seconds, the activity is reduced to one-quarter value, and so on . . .')[1]

In all the intervening years, the atom decay has never faulted him. One of his collaborators at McGill was Frederick Soddy, who discovered isotopes, i.e. 'twins' of elements which have different physical attributes but behave, chemically, like the dominant partner (of great importance when we consider U^{238} and U^{235}); and one of his students was Otto Hahn, a German, who went all the way to Montreal to work with a congenial New Zealand teacher and who was, with his discovery of uranium fission thirty years later, to confound that teacher's positive statement, 'The atom will always be a sink of energy and never a reservoir.'[2]

The trail of the atom follows Rutherford to Manchester, where another German student, Geiger (whose name is now associated with the counter, or detector of radiations), went to him and said, 'Don't you think that Marsden [a New Zealander] ought to begin a small research?' Rutherford said, 'Why not let him see if alpha

1. Edward Neville da Costa Andrade, *Rutherford and the Nature of the Atom*, Doubleday Anchor, Garden City, New York; Heinemann Educational Books, 1964, p. 111.

2. Ritchie Calder, *Profile of Science*, George Allen & Unwin, 1951.

particles can be scattered through a large angle?'[3] ('Scattered' means 'deflected'). This was something Rutherford should not have done to a young compatriot because he knew that the alpha particle was a fast, massive particle unlikely to be deflected. Marsden fired a thin beam of alpha particles at metal foil. Three days later, Geiger said to Rutherford, 'We have been able to get some alpha particles coming backwards.' Rutherford's later description of his reaction was typical. 'It was quite the most incredible event that has ever happened to me in my life. It was almost as incredible as if you had fired a fifteen-inch shell at a piece of tissue paper and it had come back and hit you!'[4]

This was one of the most significant scientific discoveries ever made. It showed that within the atom there was a prodigious repulsive force capable of deflecting an alpha particle travelling at 10,000 miles a second. But the recoil was infrequent, which meant that an immensely greater proportion of alpha particles were passing through the atom unimpeded. Following this experiment, Rutherford announced that he 'now knew what an atom looked like' and that he had an explanation for the large deflection of the alpha particle. The atom, he had decided, must consist of a very small, electrically charged, central particle in which practically all the mass of the atom was concentrated, surrounded by a sphere of electrification, very thinly spread, of opposite charge. (The diameter of the nucleus is only about a hundred-thousandth of the whole atom. The remaining space is almost empty and in this space the electrons move.)

Traffic laws in the atom

With Rutherford's experimental evidence and with help from the Quantum Theory, Niels Bohr went on to establish the fundamental laws governing the motion of electrons around the nucleus. In the first instance, the simplest subject was the hydrogen atom which has only a single positive charge on its nucleus and a single negatively charged electron. The hydrogen atom's electron can move in one of a number of possible orbits – but, under the Quantum rule, each orbit is specific or 'permitted'. The most stable of

3. Andrade, op. cit. 4. ibid.

the allowed orbits is that of lowest energy. In this track the electron keeps within one angstrom unit (one one-hundred-millionth of a centimetre) of the nucleus. Here the angular momentum (the amount of rotation of a particle in orbit) is zero. At the next higher level of energy an electron is allowed any one of four possible orbits. At a still higher energy level, the electron may have more than four allowed orbits.

Besides revolving around the nucleus, the electron also spins on its own axis, like the Earth. The electron's spin can take one of two directions: left or right. This doubles the number of permitted motions, or kinds of orbit, since in any given orbit it may spin in one direction or another. A crude analogy would be the 'dodgem' cars at a fairground, except that the cars would keep in their 'lanes' and never collide.

In an atom heavier than hydrogen, i.e. with more than one electron, the electrons' motions are much more complicated, because just as the Earth's orbit round the Sun is distorted from a perfect ellipse by the gravitational effects of other planets, so the motion of each electron round the nucleus of a heavy atom is influenced by the presence of other electrons. The Pauli Exclusion Principle, however, insists that two electrons can never exist in the same orbit of the same magnitude, direction of orbital angular momentum, and with their spins in the same direction. The 'traffic laws' in the atom are well-defined!

The Bohr–Rutherford theory of the atom in 1913 provided a 'kindergarten model' of the microcosmos. It was simple: the atom was a nuclear 'sun' with a planetary system of electrons. It disposed of the idea of atoms as 'indivisible'; it helped chemists to explain how atoms linked up in molecules; and, if the Quantum Theory gave meaning to the model, the model also gave meaning to the Quantum Theory.

For example, the single electron of hydrogen is lonely and restless. Hydrogen readily loses its negative electron and becomes a positive ion (a 'naked' nucleus): the proton. Hydrogen is highly reactive and enters easily into chemical combination with other elements. In contrast, the helium atom is extremely stable. Its nucleus (Rutherford's favourite, the alpha particle) has two positive charges which keep the two orbital electrons in tight conjunction with it. As a result, helium has not yet been observed to chemically react or combine. The same is generally true of other 'noble' gases: neon, argon, krypton, and xenon. Each is sustained by a closely knit system of electrons and does not readily part with any of them (ionize) nor ordinarily associate with other elements. Indeed the stability of the five 'noble' gases is so exceptional that their atomic numbers 2, 10, 18, 36, and 54 are known as the 'magic numbers' of the Periodic Table. It came as a great surprise to the scientific community when successful syntheses of fluorides and other compounds of xenon and krypton were announced in the 1960s.

The picture of the atom's electronic structure enabled chemists to group elements in 'families' and to predict their chemical behaviour. But physicists at the time thought it superficial or 'juggling with numbers'. The Nobel prizewinner Lord Rayleigh said, 'I have looked at it, but saw that it was no use to me.'[5] Indeed, it gave little information or clues to the structure of the nucleus. The chemical nature of an atom is determined by the number of its electrons and from that follows the number of protons. (The number must be equal in order to provide the balance of positive and negative charges.) But that does not help to explain how the nucleus itself holds together. The forces which bind the nucleus together are millions of times greater than those which bind the electrons to the nucleus.

Probing the nucleus

The idea of the various nuclei consisting of combinations of protons was not satisfactory. It was all right for hydrogen – one proton and one electron, opposite in charge. But the nucleus represents about 99·95 per cent of the mass of an atom, and this does not correspond to the sum of the mass of protons necessary to give the opposite charge to a similar number of electrons; the mass on such a calculation is excessive.

Rutherford, in 1920, made some inspired predictions in his Bakerian Lecture to the Royal Society of London. He stated that it seemed likely that a nucleus could exist having a mass of two

5. Andrade, op. cit., p. 142.

units and a charge of one unit, which would mean that it would behave chemically like hydrogen. Eleven years later, Harold C. Urey, Ferdinand G. Brickwedde, and George M. Murphy in the United States discovered just such an atom: deuterium, the isotope of hydrogen, or 'heavy hydrogen'. Rutherford also assumed the existence of a particle with a mass of three units and a charge of two units, and it materialized in later experiments as a lighter isotope of helium. But most remarkable was his anticipation of that portentous particle, the neutron. It would have no electric charge.

> It should be able to move freely through matter [he told his colleagues]. Its presence would probably be difficult to detect by spectroscope and it may be impossible to contain it in a sealed vessel. On the other hand, it should enter readily the structure of atoms, and may either unite with the nucleus or be disintegrated by its intense field, resulting possibly in the escape of a charged hydrogen atom or an electron or both.[6]

Twelve years later (1932) James Chadwick, at the Cavendish Laboratory, Cambridge, established the existence of such a particle and it was given the name 'neutron'. *This* ghost particle, which, having no electric charge, could move anywhere and enter –as Rutherford had foreseen – into the nucleus of the atom, became the 'trigger' of the atom bomb.

In those hectic days of the early thirties, in addition to the discovery of the deuteron and the neutron, Anderson discovered the 'positron', the positive electron which another of Rutherford's young men, Dirac, had theoretically postulated; Cockcroft and Walton had used their high-voltage accelerator (compounded of packing cases, biscuit tins, and glass tubes and sealed with plasticine) to split the atom; Frédéric and Irène Joliot-Curie in Paris had produced artificial radioactivity by bombarding boron and turning it into a form of nitrogen which gave off rays; they exposed uranium to a bombardment of neutrons with curious results which were to assume a more significant meaning later; Fermi, in Italy, had named the neutrino, and Yukawa, in Japan, had described the meson.

The neutrino *had* to exist, once the character of the neutron had

6. Andrade, op. cit., pp. 168–170.

been established. Inside the nucleus a neutron can live indefinitely, but when the particle is observed outside, it proves unstable. In an average time of about eighteen minutes it spontaneously ejects a beta particle (a nuclear electron) and turns into a proton. The proton and the electron are about 1·5 electron masses lighter than the neutron, so this amount, equivalent to 780,000 electron volts of energy, appears to be lost. Pauli suggested that the discrepancy might be accounted for by another particle, almost undetectable. Fermi pursued this surmise and in 1934 constructed a complete theory of beta decay. Its fundamental process is that a neutron continuously loses and regains an electron and a neutrino by emission and absorption.

Yukawa set out to describe the 'glue' which held together the protons and neutrons in the nucleus. He proposed that jointly they emitted and absorbed a nuclear field-quantum called a 'meson'. Its force would extend only over very short range and it would have a finite mass.

Three years later, a meson did materialize in the cosmic rays from outer space, detected here on Earth. It seemed to have just the properties which Yukawa had specified. It had a mass about 200 times that of the electron and was found to have positive and negative forms. But it was not what the scientists were looking for because it did not react strongly with the other particles, the protons and neutrons, and therefore could not transmit nuclear forces. Much later, in 1947, Yukawa's specifications were met by another type of meson (trapped in the emulsion of photographic plates sent up by balloon to high altitudes). C. F. Powell, a Briton, G. P. S. Occhialini, an Italian, and C. M. G. Lattes, a Brazilian, discovered that this new particle did interact strongly and had a mass of 273 electrons. This was the 'pi meson' or 'pion'.

Rutherford died in 1937, having imprinted his personality on the whole of nuclear research thus far. If at that moment a scientifically inquisitive Alice had sipped her 'Drink Me' and had slipped into the Wonderland of the nucleus, she would have found it tidily furnished with the electron (the negative beta particle); the proton; the positron (the positive electron); the neutron; the neutrino; and the still unconfirmed meson. In varying numbers and combinations, those 'elementary particles' could account for the structure of the (then) ninety-two known elements. Only a few of those elements were unstable, and their instability could be satisfactorily explained by the eccentric behaviour of the particles.

Human excitation of nuclear eccentricity had already begun.

The atom explodes

In 1919 Rutherford had shown that by using alpha particles, spontaneously released from radium, he could split the nitrogen atom, expel a hydrogen nucleus (or proton) and convert nitrogen into oxygen. He had proved what the medieval alchemists had believed, at the risk of their being broken on the rack or burned at the stake, that substances could be transmuted; he had founded modern alchemy. In 1932 Cockcroft and Walton had devised the high-voltage accelerator which, applying about 750,000 volts to a proton (the hydrogen atom stripped of its electron), used this as a projectile to hit a lithium target, split the atom, and released energy of 16 million electron volts from the individual atom. This seemed a very substantial dividend but Rutherford pooh-poohed it as a potential source of useful energy. He pointed out that only one proton projectile in ten million hits the target. He said, 'It's like trying to shoot a gnat on a dark night in the Albert Hall and using ten million rounds of ammunition on the off-chance of getting it.' Lawrence in the United States had improved on the atom smasher by devising the cyclotron in which particles magnetically directed into an endless circle could generate higher and higher velocities; i.e. higher and higher energies. But it was still 'hitting a gnat on a dark night'.

All that was changed within two years of Rutherford's death in 1937. It has been mentioned that in 1905 a young German chemist, Otto Hahn, had opted to be Rutherford's student at McGill University; it has also been mentioned that the Joliot-Curies had exposed uranium to a bombardment of neutrons with curious results. How curious emerged from Hahn's experiments (with his colleague Strassmann at the Kaiser Wilhelm Institute, Berlin) when he irradiated uranium with neutrons and found that there was transmutation into two elements, showing that the uranium had split as a result of the intervention of the neutron. Lise Meitner and her nephew, Otto Frisch, rightly interpreted this as 'fission' (a term borrowed from biology, where it refers to cell division). There was a further significant fact that the fission released a neutron which could split other uranium atoms. This could produce a possible chain reaction.

A chain reaction simply means that one event will produce another and another and another. The end result is familiar to the whole world: if a chain reaction can be sustained, if neutrons from one uranium atom can be 'captured' by the nucleus of another uranium atom, they will produce another split and the release of more neutrons. Also there will be a release of surplus energy. If the process can be sustained within a given amount of atoms (critical mass) and in an instant of time, the result will be an explosion.

It was recognized that the operative particle, spontaneously releasing neutrons, was uranium-235, which occurs in natural uranium in proportions of 1:140 of atoms of uranium-238. If enough U^{235} could be separated it would produce an explosion; but, as Fermi demonstrated in the atomic pile at the University of Chicago, it was also possible so to arrange things that the fast neutrons from U^{235} could be so slowed down that they could be captured by atoms of U^{238} to produce a man-made element, plutonium, which was unstable. This, in critical quantities, would also produce an explosion.

It thus became a question of arranging atoms in a 'lattice' so that neutrons would be captured. In a bomb the instantaneous release of energy would be catastrophic, but under control it could produce peaceful atomic energy as well as by-products, the radio-isotopes.

The next step was the H-bomb. This depends on the thermonuclear process which, as has been previously indicated, is the way by which the Sun generates and releases energy. This depends, not on splitting atoms but on fusing them. The nucleus of the hydrogen atom consists of one particle. The nucleus of the helium atom consists of four particles. If the heat is sufficient (like the 15,000,000 degrees in the heart of the Sun) four particles can be made to fuse (to become helium) with a surplus energy a million times greater than chemical energy and a thousand times greater than fission energy. Such temperatures can be produced on Earth in the instant of the explosion of a fission bomb. The trick, therefore, was to use the fission bomb as the 'percussion cap' and to surround it with susceptible material. One-plus-one-plus-one-plus-one hydrogen nuclei makes four, but there are short cuts: prefabrications in the form of double hydrogen (deuterium) and triple hydrogen (tritium) which gives two-plus-two or three-plus-one. Surplus particles are discarded in correcting the sum but so is surplus energy in abundance. As we know, the instant pressure cooker worked as the H-bomb. Another, fiercer dragon had been released. Attempts were made to tame it.

The peaceful uses of fusion energy were not as simply arrived at as were the peaceful uses of fission energy. Professor Homi J. Bhabha said, as president of the 1955 United Nations Conference on the Peaceful Uses of Atomic Energy, that if it could be tamed, thermonuclear energy would provide as much industrial power 'as there is deuterium in the seven seas'. It was then found that little was actually known about the behaviour of the particles, and to find out, yet another new science came into being: plasma physics.

(This is an interesting example of how words become corrupted in scientific usage. Plasma, to a Greek scholar, means 'mould' or 'matrix'. When it was first adopted by the biologists in such terms as 'protoplasm' [first form] it respected its origins. When, however, it became familiar to blood donors, it no longer meant 'matrix', but what went into the matrix; it meant the blood fluid without the corpuscles; but when the physicists adopted it, plasma was not a matrix nor the fluid without the corpuscles; it was the corpuscles. Plasma meant the particles separated from each other.)

'Plasma' was used to describe the Fourth State of Matter: not solid, not liquid, and not gas, but something else. If they had called it 'corpuscular flame', they might have raised the phantoms of phlogiston. But, just as it was possible by M.H.D. (magnetohydrodynamics) to extract free electrons, i.e. electric currents directly from flame, so, in plasma physics, it was sought to generate the temperatures necessary for thermonuclear fusion by the motivation of particles. This management of corpuscles was already familiar to lighting engineers in the form of strip lighting (e.g. neon signs) in which beams of electrons are passed through gases in vacuum tubes and produce luminescence. In thermonuclear applications, however, the ambitions far exceeded those of the lighting engineers in their domestication of quanta. They were seeking temperatures far, far greater than the temperature of the heart of the Sun (because, as has been already stated, the Sun is a slow oven with billions of years in which to process the fusion). Terrestrial materials cannot stand such temperatures – not metals, nor glass, nor ceramics. If, therefore, the beam of particles (e.g. deuterons) were to impinge on the walls of any transit system, the materials would disintegrate. In plasma engineering, therefore, magnetic fields, to keep the beam away from the walls, are the invisible 'wadding'.

The Fourth State of Matter is thus not just a matter of academic interest to nuclear physicists; it is a challenge and an opportunity for the technologists.

The safebreakers and the locksmiths

The safebreakers forced the lock of the atom before the locksmiths knew how it worked. This seems a disrespectful way of describing the greatest material achievement of man since our ancestors mastered fire. Moreover, the safebreakers included, directly or indirectly, almost all the nuclear locksmiths of the Western Allies, including the *doyens*, Einstein and Niels Bohr, who lent their insight and their influence. With the greatest muster of scientific brainpower ever invoked, and $2,500,000,000 worth of technology, the Manhattan Project released the energy of the atom.

As has been pointed out, Peierls and Frisch, in their report to the British MAUD Committee, were able to say that a chain reaction could be sustained and a superbomb produced. Hahn and Strassmann had provided the clue. Meitner and Frisch had 'read the message' in terms of fission. Bohr and Wheeler had proposed their 'liquid-drop' model of the nucleus.

The 'liquid drop' was an ingenious evasion of all the unanswered questions. It was recognized that the forces which bound the nucleus were millions of times greater than those that bound the electrons to the nucleus. It was not known how those forces were created, and even if that had been entirely known, the scientists would still have been confronted with the prodigious difficulties of calculating the results of those forces upon a large number of protons and neutrons which interact with one another strongly and at extremely short distances. The Bohr–Wheeler version simply treated the nucleus as though it were a drop. Just as in a water molecule the chemical identities of hydrogen and oxygen are merged, so in the nuclear 'drop' the protons and neutrons lost their identities. And the 'drop' lent itself nicely to the concept of fission: it could elongate like a raindrop on a window-pane, with a waistline which diminished to a snapping point. Frisch had had the same sort of borrowed imagery when he thought of the dividing biological cell.

Needless to say, the mathematics and physics of the Bohr–Wheeler 'drop' were not as simple as the image. How the invading particle, the neutron, upset the homogenous proton–neutron 'drop' and caused it to divide and how in that division spare neutrons were released to repeat the process, were matters of laborious and ingenious calculation and experiment. Nevertheless, this was high-level know-how which was ultimately translated into technological know-how – the separation of U^{235}, the assembly of the lattice of uranium and graphite which slowed down the fast neutrons and contrived their capture by U^{238} – to produce the predictable plutonium and the ultimate 'weaponry' by which nonexplosive fractions of fissile material could be suddenly massed together into a concentration which would flash off as energy.

With deference, however, one would repeat that they had cracked a lock of which they did not know the combination, nor how the wards fitted into place. As a reminder: they were then conceiving the nucleus as consisting of protons, neutrons, beta particles, gamma rays ('hard' X-rays) and Yukawa's 'glue', the still-unconfirmed gripping mesons.

J. Robert Oppenheimer, as head of the laboratory at Los Alamos, assembled all his colleagues' knowledge, brought key men together in the actual operations, and compounded knowledge, experience, and material into the bomb which exploded at Alamogordo, New Mexico, on that morning which changed world history. In 1958, when, with Niels Bohr, he officiated at the opening of the Institute of Nuclear Science at the Weizmann Institute of Science, Israel, Oppenheimer said, 'If you had asked me ten years ago what the structure of the nucleus looked like, I might have told you. Ask me ten years from now, and I may be able to tell you. At the moment, I cannot tell you.'

Between July 1945 and 1958, the tremendous impetus to nuclear research which the nuclear bomb had produced had increased the known components of the nucleus to thirty. Scientists still referred to them as 'elementary particles', and though they were detectable only individually, they could nevertheless conceive them as forming an orderly pattern and a consistent relationship. Within the next five years, seventy other subatomic objects had been discovered and, with this embarrassment of riches, the scientists began to paraphrase George Orwell's *Animal Farm* statement: 'All animals are equal, but some animals are more equal than others.' All nuclear particles are elementary but some must be more elementary than others!

In 1952, Professor Erwin Schrödinger was already saying:

> Fifty years ago science seemed on the road to a clear-cut answer to the ancient question 'What is matter?' It looked as though matter would be reduced at last to its ultimate building-blocks – to certain microscopic but nevertheless tangible and measurable particles. But it proved to be less simple than that. Today a physicist can no longer distinguish significantly between matter and something else. We no longer contrast matter with forces or fields as different entities; we know that those concepts must be merged.

What has happened had happened before in physics. The thinking of the previous fifty years had become 'classical'. It had been adequate for perceiving order in a limited number of observations, but the efforts to adjust the new observations to the paradigm had become cumbersome. It was like packing at the end of a tourist trip; the bought presents just would not fit into the suitcase which Rutherford, Bohr, and others had provided.

This was what happened to the Ptolemaic System.[7] The system

7. Cf. p. 34.

served its purpose for centuries and still is useful for working approximations, but when generations of astronomers tried to make Ptolemy's 'laws' of the epicycles, describing the motions of the planets, apply to what they actually observed, the tradition became so cumbered by contradictions that it just could not be true to nature. Only the Copernican Revolution could banish the contradictions. The same was true when the spectroscopists, at the beginning of the twentieth century, tried to apply classical electrodynamics. They were not presumptuous; they wanted to work within the paradigm. But it was like amending the tax laws; every time a flagrant breach was closed, another would open. Studying the light emitted by excited atoms, they found a profusion of discrete wavelengths that were at total variance with the 'authorized' wavelengths of the classical theory. They accumulated so much empirical information, including sets of 'selection rules' governing the permissible states of excited atoms, that it became necessary and possible for Werner Heisenberg, Erwin Schrödinger, and others to formulate quantum mechanics, capable of predicting most of the states of matter on an atomic and molecular scale.

Strange particles

To see what had happened to cause physicists to substitute adjectives such as 'strange' for 'fundamental' as applied to particles, we have to distinguish between 'atomic' and 'nuclear'. By the middle of the 1930s the Theory of the Atom was essentially complete. The properties of ordinary matter could be mathematically deduced, in an entirely satisfactory way, in terms of negatively charged electrons around positively charged nuclei. Most of the problems with which physics and chemistry had struggled for centuries were for all practical purposes solved. In the process, 'pure' chemistry had merged into physics, because chemical phenomena could now be explained in terms of physical interactions.

The difficulties began when the scientists turned from the 'atom' to its 'nucleus'. They knew that the nucleus was made up of protons and neutrons – at least, their sum accounted for the mass. When, however, they began to observe nuclei, smashed either by cosmic rays or in machines, they found that entirely new types of matter were created: a bewildering variety of short-lived particles which apparently do not exist within atoms of ordinary material. Instead of a theatre-in-the-round, with a well-rehearsed cast of four or five characters, they found themselves staging a Hollywood spectacular with performers of whom the casting director had never heard and who danced in briefly out of nowhere.

Some physicists even began to raise heretical questions such as 'Are these particles really part of any pattern of nature or are they the artifacts of our own invention?' The question is not superficial. We say, 'Here is a light bulb.' But the bulb contains no light. We say, 'Ah, but there is electricity and when we switch it on, the filament produces light.' But, intrinsically, the electron particle has no 'light'. The molecules of tungsten in the filament have no light. We send the electrons, artificially generated, through non-light copper wires to excite the atoms of tungsten so that the electrons of the tungsten atoms jump from one orbit to another and in the process release photons. The photon is a quantum of radiant energy. It always travels with the velocity of light. It can never be at rest. It possesses mass only by virtue of motion. By a man-manipulated process, from the generating station to the filament, we have released photons from an element which (unlike the radium of our luminous wristwatches) does not, in Nature, emit light quanta. The light bulb, therefore, has no intrinsic light – only what we create by external intervention.

Are the experimental physicists 'inventing' particles in the same way that Swan and Edison invented light bulbs? The scientifically reassuring answer might be that many of the particles are first observed in cosmic-ray research and then efforts are made to reproduce them; or, if they are found in terrestrial experiments, efforts are made to find them in cosmic rays. Cosmic rays are natural, and so this is a double check with Nature. But, the scientific heretics would say, the methods of detection are man-made. In detecting, we may be creating.

Without endorsing the 'heresy', let us look briefly at some

methods of detection. Firstly, no one has ever seen an atom, or a nucleus, or a nuclear particle. When crystallographers produce those beautiful and convincing models with coloured beads or billiard balls and say, 'This is how the atoms of the various elements are arranged in a molecule of penicillin,' what you are seeing is not what they see. They use X-rays, and see not atoms, but the aura of atoms in the form of diffractions which, being spectrally characteristic for each element, reveal the presence. Moreover, when a nuclear physicist gives the precise particulars of a particle ('A proton has a mass 1,836·1 times that of an electron and has a charge of plus one') it is like weighing and measuring Wells' Invisible Man, whose presence was revealed when he left the footprints of his unseen body in the snow.

The 'footprinting' of nuclear particles began with an inspiration which came to C. T. Wilson, Rutherford's colleague, when he was walking through his native Scottish mist. He conceived the 'cloud chamber', in the fine mist of which a charged particle would leave an ionized trail – like that left in the sky by a high-speed jet. According to how far the visible track of a particle travelled, or whether it turned left or right, or what happened when it collided with other nuclei, a great deal could be learned about the identity of the particle. Another identification device was provided by Geiger with his detector by which a particle would cause an electrical discharge and reveal itself as a 'click'. Those earlier devices have been refined and methods have multiplied: scintillometers, to make Geiger's clicks visible; computerized counters to replace the 'mileometer' of Cockcroft and Walton's first atom-splitting machine; bubble chambers; highly sophisticated emulsions for the photographic plates which Powell first sent up by balloon to trap incoming cosmic rays; space vehicles which with their instruments can intercept and identify particles and rays and telemeter the precise identification back to the 'Interpol' of physics. The versatility and the precision of the detectors are impressive; the information they disclose must surely be exact, i.e. true. 'But,' say the sceptics, 'is what we are measuring the intrinsic or extrinsic nature of the particle? Are we identifying the man who came in out of the snow or a snowman we have built up out of a snowball?'

The nuclear ghetto

This is the age of the linear accelerators and the '-trons'. The first derives from Cockroft and Walton's proton cannon and the second from E. O. Lawrence's cyclotron. Nowadays, the first can be over two miles long and the second bigger than a race track. To call them 'atom smashers' is a misnomer; rather, they are 'atom makers'. And to say that they make snowmen out of snowballs would be provocative. But the analogy without the implication has a relevance. And for it we go back once again to Einstein's $E = mc^2$. If one has energy in the electron-volt sense, one can produce mass; and the greater the velocity at which a particle travels the greater the mass. One can therefore tailor-make particles, i.e. energy identifiable as matter. In fact, mass can be expressed in terms of electron volts. For example, when, fulfilling Yukawa's theoretical prediction, the pi meson was detected by the cosmic-ray experts and was found to have a mass 273 times that of the electron, a particle with an energy of 137 million electron volts (Mev) was reproduced in a machine. In that case, and in many others, it was nice to reproduce something which fitted into a theory – like Adams and Leverrier independently finding the planet Neptune with the help of Newton's laws. But are some of the multiplying particles man-made products of the process of manufacture? They do not seem to fit into any theory. They are like gatecrashers at a table-set dinner party, and some physicists, despairing of ever restoring protocol, say in effect, 'Why have a set table? Why not have a cocktail party?' In other words, must there be a 'law'?

As Schrödinger said,

> Physics stands at a grave crisis of ideas. In face of this crisis, many maintain that no objective picture of reality is possible. However, the optimists among us (of whom I count myself one) look upon this view as philosophical extravagance born of despair. We hope that the present fluctuations of thinking are only the indications of an upheaval in old beliefs which in the end will lead to something better than the mess of formulas which today surround our subject.

The optimists persist, in spite of the aggregation of discoveries which confuse, and they search for order and uniformity in the microcosmic universe of the nucleus, without which the Unity of Matter, embracing the macrocosmos, is untenable.

Perhaps it will be possible to find, on the analogy of the Quantum Ladder, a Jacob's Ladder reaching from the heart of the nucleus to outermost space. The Quantum Ladder separated Nature into levels which could be considered separately but in ultimate relationship to each other. It applied admirably to the electronic structure of the atom, by treating it as though it were an apartment house.

The ground floor, or orbit of lowest energy, had only a single apartment. Two electrons of opposite spin could occupy it. The second floor was split-level, with four apartments – one slightly lower than the others. Each of the four apartments could again be occupied by two electrons of opposite spin. And so on, floor by floor (the floor representing the energy level in terms of angular momentum, the speed with which an electron moves round its orbit), with a prescribed number of apartments, each for occupancy by a pair of electrons. It happens that atoms are most stable when all the apartments are fully occupied. That is why the electron of the hydrogen atom, solitary in its apartment on the ground floor, is restless. It picks up a roommate, i.e. enters into chemical relationship with another element. Or it quits, leaving the hydrogen as a positive ion, the proton. For an electron to move from one quantum level to another is a significant event. It releases quanta (or packets) of energy. (The orderly promotion of such shifts provides us with transistors.) On the analogy of the apartment house the nucleus has become a veritable ghetto.

The object of the nuclear theorists is to restore order by inventing something analogous to quantum mechanics: a unified symmetry which would account for all the phenomena, and against which any aberrations could be tested. It means sorting out the experimental data into patterns. The difficulties are enormous.

Einstein tried, and admitted failure, to produce a theory unifying gravity and electromagnetism, which would cover the forces in the nucleus. He could be excused because it is now evident that there are recognizably four different kinds of forces operating in the nucleus. The so-called nuclear force is revealed by the strongly interacting particles. The second force is electromagnetic. The third force is called the 'weak force', revealed by the weakly interacting particles. But the weakest of all is gravitational force.

The strongly interacting particles, such as the proton and the neutron, are clearly distinguished from other particles. They interact through the strong, short-range (100,000,000,000,000th of a centimetre) nuclear force. All particles seem to participate in this strong interaction, with the conspicuous exceptions of the four particles (called leptons) which are the electron, the muon (or mu meson), and the (now) two kinds of neutrinos.

None of the strongly interacting particles has a small rest mass, i.e. the mass that the particle would have if it were motionless, which is the lowest mass a particle can have. The lightest possible interacting particle is the pion (or pi meson) with a mass equivalent to an energy of 137 million electron volts compared with the half-million electron volts of the electron's mass, and compared with the negligible mass of the photon and neutrinos.

One thing is known which is suggestive of the apartment house of electrons in the outer atom, and that is that there can be different energy levels indicating different degrees of binding energy between the neutron and protons.

This 'strong' nuclear force is a hundred times greater than the electromagnetic force which, in turn, is a million million times greater than the 'weak force'. The weak force operates at very short range but is stronger than gravitational force, which is 10^{-39} of the strength of the nuclear force.

Since two of those forces are peculiar to the microcosmic universe of the atom and cannot be related to falling apples or spinning planets or copper wire wrapped round iron quoits, it is obvious that there is little to be borrowed, or eschewed, from previous experience. What we are looking for is a parthenogenic paradigm born of the nucleus itself.

For one thing, it will have to encompass anti-matter. Dirac theoretically postulated (1928) the existence of a positive electron, and Anderson experimentally delivered the 'positron'. (If a positron meets an electron they both vanish without trace.) More and more anti-particles were shown to exist and to suggest that everything in Nature had its opposite.

The overthrow of parity

While the physicists were rounding up maverick particles and trying to fit anti-matter into the nuclear corral, they were confronted by a new difficulty: the overthrow of parity.

One of the solid pillars of modern physics was Leibniz's 'great principle': 'Two states indiscernible from each other are the same state.' It supported the Theory of Relativity and the laws of conservation of energy, momentum, and so on, upon which the understanding of Nature is built.

The important word is 'indiscernible'. Among the indiscernibles are absolute space, time, and direction. As Philip Morrison put it,

> Think of the conventional world map. To each place are assigned a latitude and a longitude – a pair of numbers. The numbers are of great utility and convenience, but they are in no sense attributes of the places; they have no physical significance. If the starting point for counting were to be shifted from Greenwich to Timbuktu, the numbers would change but no mountains would be moved. The numbers are merely arbitrary labels. And this is the manner in which space in general is treated in physics. The coordinates specifying positions in space describe only relative positions. We try to formulate our physical laws by the use of mathematical schemes in which absolute positions in space never enter. Whatever our frame of reference, we say, space remains invariant.[8]

Einstein's special theory of relativity depends on the indiscernibility of absolute coordinates. The fact that physical equations cannot refer to absolute time, space, or orientation is the mathematical basis for the classical laws of conservation. One of those 'laws' was the conservation of parity, which rests upon the taken-for-granted indiscernibility of right–left. In other words, the 'mirror image' of a process is indistinguishable from the actuality.

The principle of parity (right–left) insists that for every process there exists its exact reverse counterpart and that, for example, a molecule would function exactly the same way if it were mirrored. We talk about North and South Poles in magnetism and apply conventions, like colouring the compass-needle tip which points

8. Philip Morrison, 'The Overthrow of Parity', *Scientific American*, April 1957, p. 47.

to what we call 'north', but there is no intrinsic physical distinction between the North Magnetic Pole and the South Magnetic Pole. There is nothing in the laws of electromagnetic fields that permits an absolute distinction between right and left, any more than there is a natural law which compels our feet to walk on one side of the road rather than the other; our conventions, expressed in traffic rules, decide that.

The conservation of parity was a built-in piece of the furniture of mathematics and physics. Then a couple of those 'strange' particles came along and spoiled everything. Parity was toppled over.

It began as the 'tau-theta puzzle'. There were two mesons called tau and theta. Tau disintegrated into three pi mesons. Theta disintegrated into two pi mesons. But, in every other property except the mode of decay tau and theta were identical and could have been considered to be the same particles but for the fact that tau decayed to a set of pi mesons (pions) of odd parity and theta to pions of even parity. That, said the 'law', makes them distinct particles.

Tsung-Dao Lee, of Columbia University, and Chen Ning Yang, of Princeton Institute for Advanced Study, challenged the law. Maybe it did not apply to the realm of weak interactions in which the tau-theta decay took place. They decided that there should be an experiment to test whether right and left could or could not be distinguished in the subatomic realm. Tau and theta, which had started all this, were not suitable subjects for experiment; each has a lifetime of about a billionth of a second. But there was a candidate, the beta electron of the nucleus that, unlike its electromagnetic namesake which is external to the nucleus, was one of the particles in the realm of weak interactions. Radioactive cobalt, which emits a beta electron, has a half-life of 5·3 years. The experiment suggested involved lining up the spins of beta-emitting nuclei along the same axis and then seeing whether the beta particles were emitted preferentially right or left along the axis.

It took a team six months to design, prepare, and set up an experiment which lasted fifteen minutes. In that time, the beta particles emitted from the lined-up cobalt nuclei showed an asymmetry relative to the direction of the axis.

This proved that in the realm of weak interactions parity did not prevail. On this evidence, repeated and followed up since, one of the great invariance principles of science was undermined. It could not be sustained in the realm of weak interactions. That does not wipe parity off the blackboard and there is no reason to believe that it does not still hold as an invariant in the macrocosmos. But the evidence destroyed its universality. It also raised questions as to whether the other great 'truths', such as the conservation of energy and the conservation of momentum, regarded as self-evident and inalienable, were not similarly dubious in the microcosmos of the nucleus.

Pessimists saw in all this a confusion worse confounded and began to ask themselves whether uniformity was necessary, let alone attainable. Might not Heisenberg's Uncertainty Principle, which recognizes that the behaviour of individual particles is unobservable and unpredictable, go even further and confer a uniqueness on every nuclear phenomenon?

Optimists, on the other hand, regarded the overthrow of parity merely as evidence that we had been exceedingly naïve and that all it showed was that there were more things in the heavens and in the nucleus than were dreamt of (thus far) in our philosophy. No one until Tsung-Dao Lee and Chen Ning Yang had asked the right question because the answer was taken for granted. Liebniz had said it – 'Two states indiscernible from each other are the same state' – and physicists had shown that the 'mirror image' held true in visible Nature, and the mathematicians had evolved the parity concept. Why not start again and, pursuing the questions, seek answers which might provide the bridge between the microphysics of particles, and cosmology, the physics of great distances?

A ball game without a ball

In theoretical physics, one can have a ball game without the ball. To the uninstructed onlooker, it is all very confusing. There do not seem to be any rules, because, apparently, none of the participants are obeying any. An imaginary ball is delivered with great force, but, as far as the onlooker is concerned, it could be a table tennis ball, a baseball, or for that matter, a cannonball. The imaginary ball is hit with an imaginary bat and everyone goes chasing around to catch it, or, more confusingly, to reach an imaginary base without any reference to the imaginary ball. Someone is keeping score, but of what?

There are rules, of course, because the theoreticians made them and they are so familiar with them that, like a well-informed ball game fan, following a report on radio, they can tell what is happening without actually watching. If the ball is of the size they have specified (the mass of a particle) and if it is projected at a specified speed and it curves to the right or left (charge, positive and negative) it will recoil from the bat at a prescribed speed and in a prescribed direction, to finish up at a prescribed place. If all those things happen, it is all according to the rules and they know that their mathematical concepts were right.

The experimental physicist on the other hand wants to see what is happening. The ball, by virtue of its submicroscopic size, is still invisible, but his job is to identify it as a particular kind of particle, assess its mass, check its electric charge, measure its force and direction, and see whether it is behaving according to the theoreticians' rules or whether the rules have to be altered. Or, where the rules prescribe a particle not hitherto identified, he may find it and then everyone is happy – the theoretician because his idea has been proved right, and the experimental physicist because he has given it reality.

In the Alice-in-Wonderland of the nucleus, the theoretical physicist can do without the ball, imaginary or otherwise. The 'something' which is passing backwards and forwards on the pitch can be explained, in his mathematical terms, as an interaction between the forces of the nucleus. He is prepared to 'go along' with the nuclear physicist and see how fundamental particles can be described in terms of those interactions to the point of identification. In effect, he says, 'If you want your ball, you can have it.' But the two positions, that of the theoretical physicist and that of the experimental physicist, must be reconcilable and not flatly contradictory.

There is a theory of elementary particles which is generally accepted, but only for lack of a workable alternative. It is the 'quantum-field theory'; both Einstein and Dirac quarrelled with it but could not replace it. The main objection is that it fails to give a full explanation of the 'why' of elementary particles. It does not claim to do so; it is descriptive and not explanatory.

Its function is comparable to that of chemistry before the beginning of the twentieth century – which was to describe the properties of chemicals and their interactions, not why they existed. That had to await atomic physics. And so the quantum-field theory treats elementary particles just as the chemists treated the elements. It starts from a list of elementary particles, with specified masses, spins, charges, and characteristic interactions one with another. From this information, it is possible to deduce what will happen if particle A collides with particle B.

Although modern physicists would argue that a classical field is only a large-scale manifestation of a quantum field, the difference of definition helps in the understanding of what is meant by 'field'. A classical field is a kind of tension or stress which can exist in empty space in the absence of matter. (The 'empty' is important because the field theory upset the concept of ether as a sort of sea in which waves could be propagated; in other words, the 'field' assumes the waves without the ether.) The tension reveals itself by producing forces which can act on any material objects that happen to lie in the space which the field occupies. The standard examples of classical fields are the electric and magnetic fields which push and pull electrically charged objects and magnetized objects. Faraday showed that these two fields exerted effects on each other and that a changing magnetic field produced electric forces; hence the electricity generators of today. Clerk Maxwell went further and formulated mathematically the 'laws' of behaviour of electric and magnetic fields. He found that in any space where a changing magnetic field exists, an electric field must also exist and vice versa. The characteristic mathematical property of a classical field is that it is an undefined 'something' which exists throughout a volume of space and which is defined by sets of numbers, each set denoting the field strength and direction at a single point in space. Maxwell went further still and deduced that electric and magnetic fields could exist not only near charges and magnets but also in free space entirely disconnected from material objects. His guess was that in empty space such fields would travel with the velocity of light and, ingeniously, that light consists of travelling electromagnetic fields. How right he was is demonstrated every hour on the hour and throughout the hour by our broadcasting stations. They manufacture travelling electromagnetic fields and send them out as radio waves.

Another classical field is gravitation. It acts on all material objects in a given region of space. It keeps the solar system together, but it is, in laboratory terms, a weak field, because it is difficult to produce, in eloquent proportions, on a miniature scale. Thus no freely travelling gravitational waves, comparable to the electromagnetic ones, have declared themselves, although they would almost certainly exist in the neighbourhood of a rapidly oscillating mass of sufficient size. Nor has it been possible to measure any possible interactions between the gravitational and electromagnetic fields. Einstein applied his Theory of Relativity to the electromagnetic field, and in 1916 he extended it to the gravitational field. Separately, they have stood the test of time, but he failed to produce a satisfactory unified theory.

The classical field theories, as updated by Einstein, give an entirely satisfactory explanation of the behaviour of large-scale physical phenomena. No one seriously doubts that the field theories are correct as long as we are discussing any object much bigger and heavier than a single atom. But they fail completely to describe the behaviour of individual atoms and particles.

This to the constitutionalists seeking the Unity of Matter is as dismaying as the Volstead Act was, in another context when, seeking only sober, orderly behaviour, it produced the law-breaking anarchy under Prohibition. Laws more often broken than respected have to be replaced. To cope with the non-compliance of the atomic particles, the physicists had to invent quantum mechanics and the idea of the quantum field.

A basic axiom of the quantum mechanics is, as has been repeatedly mentioned, the Uncertainty Principle. The behaviour of an individual electron or particle is never precisely predictable and quantum mechanics can reveal only how 'populations' of particles, not individuals, will fluctuate, on an average and over a long period of time. The fluctuations are not observable when examined by instruments which served in the classical field. Only when the effects of an electromagnetic field are applied to a single atom do the quantum fluctuations of the field become noticeable. The Lamb–Retherford experiment of 1947 was one of those scientifically satisfying exercises which, using the cavity magnetron (devised as an airborne power unit for radar during the Second World War) to observe the effects of electromagnetic fields on single hydrogen atoms and by techniques of microwave microscopy, established measurements which coincided with the requirements of the quantum-field theory.

These quantum-field fluctuations, mathematically expressed, give us the ball game without the ball. The rules are sufficient to explain the course of the game.

This is because, in quantum mechanics, energy in a quantum field can exist only in discrete units: the quanta. When the theory is applied to the quanta in detail it is found with satisfying regularity that they have the properties of elementary particles, which had been observed. But different quantum fields exist. Each has its own properties and each fills the whole of space. Between various sorts of fields, various kinds of interactions are noted. The structure of matter may depend on the permutations and combinations of those particles and the variations which the interaction of the fields produces.

Embarrassment of riches

So far, so good. Immediately after the Second World War, the mathematicians, the theoretical physicists, and the nuclear physicists were still on speaking terms. Quantum mechanics and the field theories were holding their own. The experimental physicists were coming up with the answers which fitted and they themselves were now the satraps of science. Having proved their capabilities by unleashing the atom bomb, they could ask for, and get, anything they wanted in the way of equipment for the study of high-energy particles. Out of those machines came new particles. Particles? The query is justified because, remember, they were seeing and measuring *effects* with increasing refinement and accuracy, but what produced those effects – particles? waves? or 'wavicles', an interaction, not necessarily of substances but of forces?

To grasp what they are actually seeing in their cloud chambers and bubble chambers, we can find an analogy, a very crude one, in another ball game: football. Wells's Invisible Man is charging through a swarm of invisible players. You can see his path – the ionized track – but you cannot see him. He collides with other players and weird things happen. He may cause one of the players to become two other players, reeling off separately on their own ionized tracks to collide with others and disintegrate them. He may disintegrate himself or he may even, in full career, change his shirt. (For instance, take a collision between two protons: usually one of the protons uses the collision energy to change into a neutron plus a positive pi meson, but it might avail itself of the collision energy to become a neutron, plus a positron, plus a neutrino.) But all kinds of incidental things happen: players change their direction – become positive instead of negative. They may even seem to change sides – become anti-particles. They may lose their identity, i.e. reveal no track because they have become neutral particles, with no electric charge to ionize a trail, and yet still manifest themselves in collisions which produce other particles which reveal themselves as tracks.

All this is happening under artificial conditions (in the chamber or in the emulsion of photographic plates), whether the 'Invisible Man' is a cosmic-ray particle supplied by Nature or a particle manufactured by the high-energy machines.

There is nothing wrong in this. Scientists have to control the conditions in which they make their observations and measurements, otherwise they cannot check what they are doing and are just seeing haphazard events. If they are to study the substructures of matter, they must study them at different levels. To break a structure into a substructure, they have to supply energy (appropriate for the different levels), and they can measure the energy they have put in to produce the particle which appears. (For example, if 200 million electron volts are put into a proton, a meson is knocked out with a mass-energy equivalent of 137 million volts, and they know how to account for the discount.)

But, just as the biologist, when he has carried out experiments *in vitro* (in the test tube), has to ask himself, 'But is this actually how it happens *in vivo* [in the living organism]?' so the nuclear physicist has to ask himself, 'Are those effects which I am observing in man-created conditions really what happens in Nature?' And there is the quandary, which became aggravated from about 1948 onwards. He found himself with an embarrassing abundance of particles, each of them individually fascinating. By the late 1960s, there were over a hundred, with new ones averaging two a month. He had to sort out the grain from the chaff, and he was never quite sure whether the particle he was observing was not, like puffed rice, an artifact he had shot from a cannon.

What they were looking for was a theory which the descriptive quantum mechanics did not securely provide, that would put everything in logical order so that the extraneous effects could be discarded or explained as incidents in defined process.

One approach was 'bootstrap physics' proposed by Geoffrey F. Chew and S. C. Frautschi of the University of California, at Berkeley. It is not as 'folksy' as it sounds. To quote an authorized version:

> According to the bootstrap hypothesis each strongly interacting particle is assumed to be a bound state of those channels with which it communicates, owing its existence entirely to forces associated with the exchange of strongly interacting particles that communicate with crossed channels. Each of these latter particles in turn owes its existence to a set of forces to which the first particle makes a contribution. In other words, each particle helps to generate other particles, which in turn generate it. In this circular and violently nonlinear situation it is possible to imagine that no free, or arbitrary, variables appear (except for something to establish the energy scale) and that the only self-consistent set of particles is the one found in Nature.[9]

This has been called 'the democratic approach to elementary particles'. It says, in effect, that it is not very sensible to talk about 'elementary particles' and then try to construct every other particle from those basic particles. Rather, it should be possible to take any three particles in Nature and combine them

9. G. F. Chew, M. Gell-Mann, and A. H. Rosenfeld, 'Strongly Interacting Particles', *Scientific American*, February, 1964 p. 93.

properly so that they themselves will form others. These mutually propagating particles, interacting and exchanging their forces, would produce a consistent picture of all matter.

Another approach to the superfluity problem is classification. Linnaeus did it with plants. Nature seemed to have produced plants in indiscriminate profusion and confusion until he began to arrange them by their characteristics and identify them by categories and show that in their apparent diversity there was a consistency. Another and closer analogy would be the Periodic Table. Indeed when Mendeleev came to their rescue, the chemists of the nineteenth century were in much the same predicament as the nuclear physicists. He labelled the elements by their known individual characteristics and then grouped them according to the characteristics they showed. The pattern was so clear (showing how orderly Nature was) that if there was a gap in the table – an element the existence of which had not yet been revealed – the chemists could confidently look for (and find) an element answering that precise description.

Such a classification of the ill-sorted particles was independently proposed by Murray Gell-Mann of Berkeley and Y. Ne'eman, a colonel of the Israeli Army engineers. Each suggested a unified system of symmetries. (Symmetry is one of the easier-to-understand concepts. It simply assumes that everything is arranged in a balanced pattern – china- dogs at each end of the mantelpiece – and that any item which does not fit into that pattern is bogus, and that any item missing from the pattern has to be found.)

The Eightfold Way

This system of symmetries was picturesquely called 'the Eightfold Way', recalling Buddha's injunction:

> Now this, O monks, is noble truth that leads to the cessation of pain: this is the noble *Eightfold Way*, namely, right views, right intention, right speech, right action, right living, right effort, right mindfulness, right concentration.[10]

Poetic licence was being applied to physical right-and-left rather

10. ibid., p. 89.

than moral right-and-wrong: the system involved the operation of eight quantum numbers.

The 'Eightfold Way' assumed that the particles which respond to the strongest of the four natural forces in the nucleus are not by definition 'elementary', but exist only as components one of another ('multiplets'). These multiplets would be consistent combinations identifiable by their properties, such as charge. The mathematics of the system was derived from an algebra evolved in the nineteenth century by the Norwegian, Sophus Lie. The simplest Lie algebra involves the relation of three components, each of which is a symmetry operation consistent with quantum mechanics. But there is also a Lie group SU_2 which means 'special unitary groups for arrays of size 2×2, subject to the condition that the number of independent components is reduced from 4 to 3'. (Hence, the 'special', which is the mathematician's way of saying that two times two does not necessarily make four.)

The 'Eightfold Way', however, relies on SU_3, which stands for 'special unitary group for arrays of size 3×3', again with the provision that three times three is not nine, but eight. (Actually the accounting is much more complicated than this simplification suggests.)

This 'symmetry game', as Gell-Mann called it, produced an immediate result. It predicted a new particle needed to fill out a multiplet of ten components. The specification for the missing particle was precise: it would be a negative omega baryon (a heavy particle) at about 1,532 million electron volts. It would live about 100-billionths of a second. The Big Machines duly delivered it.

Although the SU_3 system systematized particles according to charge (an 'internal' quality, according to the physicists), it did not provide for 'external' qualities such as spin. This flaw produced another theory based on SU_6, proposed by Radicati, an Italian, Gursey, a Turk, and, independently, Sakita, a Japanese. This arose from a suggestion by Eugene Wigner, the Nobel prize-winner in 1963, to deal with atomic nuclei. SU_6 grouped multiplets together into supermultiplets and took care of the internal qualities (charges) and the external (spin).

Again, it was not adequate. Wigner had been dealing with

atomic particles which did not move at high speed, and SU_6 could, for his purposes, consider particles as though they were at rest. The later theory had accepted this. Subatomic particles, however, travel at terrific speeds and the calculations affecting them should have included (and did not) Einstein's Theory of Relativity.

In 1965, Abdus Salam proposed SU_{12}. This Pakistani, Professor at Imperial College of Science and Technology, London, and Director of the International Atomic Energy Agency's Theoretical Physics Centre, Trieste, recalled an elegant but esoteric suggestion which his former tutor Nicholas Kemmer had made in 1939. The result described, at the same time, the internal and external properties and also how the particles ought to behave if they had any respect for Einstein. Even this vade-mecum of the subatomic universe had its shortcomings.

Any unified system should explain the relative strength of the forces within the nucleus: strong (named 'hadron'), electromagnetic, weak, and gravitational. (To measure the last, the experimentalists would have to show whether a particle fell down or fell up, and one of the things they sought to do, to find this, was to reduce the speed of the electron from thousands of miles a second to tens of miles a second.) It should relate the strength of those forces to mass. It should account for the 'fermions' and 'bosons'. The fermions cannot be at the same place at the same time, but the bosons are together.

The basic trouble is environmental. Just as man is at one and the same time an individual entity and a creature of his environment, so the elementary particles (if we can use the term any longer) have inherent properties and interacting properties. The Theory of Relativity makes it clear that there are intrinsic properties which cannot be separated from external properties. There is no objective way of separating the intrinsic from the external, and that is the limiting factor of experimental physics. The particle at which they are looking is the 'here-and-now' and they identify it as what they measure.

Is it a man covered with snow, or a snowman? Only a theory which fits all the facts, and at the same time discards the artifacts, can resolve the dilemma.

'Manifestations'

Sir Cyril Hinshelwood, O.M., F.R.S., suggested in his presidential address to the British Association for the Advancement of Science (1965) that 'manifestations' would be more appropriate than 'particles' as a definition of the phenomena which the nuclear physicists are observing and measuring. Certainly their nature and behaviour are quite different from anything observable in macrocosmic experience. The energies are many millions of times greater than those found in chemistry. As discovery followed discovery, the subatomic picture became more and more bewildering. If 'they' were treated as particles, then some were heavy, some light, some stable, and some ephemeral. They appeared and disappeared. They differed in mass and charge, in duration, and in the intensity with which they interacted one with another. Forces existed which were neither electrical nor gravitational and apparently had no counterparts in tangible experience. The nonelectrical forces were extremely puzzling. Some were fantastically powerful and some no more than a nod or a wink in particular relationships. While the experimentalists tried to measure them, the mathematicians expressed them in their poetry of symbolism and some sort of consistency began to emerge. It was Pythagorean in its geometrical symmetry. In their search for a fundamental law, they were finding, if not unity, at least uniformity. Reinforced by mathematical abstractions, the experimentalists were beginning to see the relevance of their 'manifestations', one to another.

13 Matter of Opinion

'I think therefore I am'

The ultimate aim of the scientist is to provide a logical explanation and a numerical definition for everything from the egregious behaviour of the latest subatomic particle to the uttermost quasar; to explain the supernatural in explicit terms of the natural (i.e. measurable); to reduce the occult to the substantial and, in the process, to account for life itself without recourse to special creation or *élan vital.* Some would say that this is a matter of opinion. The reductionists would say that opinion itself is matter.

Apart from theological arguments, or the conflicts of science and religion, the idea of the continuity of matter and life – inanimate into animate – has always presented philosophical difficulties. The reason is obvious: it is a living being who asks the questions. Descartes (1596–1650) expressed it as *cogito ergo sum* ('I think therefore I am') and Spinoza (1632–77) redefined it as *ego sum cogitans* ('I, being conscious, am existent'). Thus science, which presumes to make everything accountable to reason, itself exists only in the realm of ideas, in the human mind, and *mind* is something different from that conglomeration of cells and atoms which is observable and measurable as the *brain.*

Descartes, in substituting a logical system of mathematics for the collection of curiosities and occult qualities which had passed as medieval science, acknowledged and accepted the dilemma. His mathematical reasoning employed *intuition, deduction,* and *enumeration.* Intuition is the self-evident concept which one proceeds to examine. Deduction is analysis and synthesis – converting the intuitive idea into manageable terms and reassembling the terms into a stable model of the idea. Enumeration is the expression of the various steps in the mental process in a mathematical shorthand which other mathematicians can read and repeat with the predictable certainty that the result each time will be the same. For this purpose there must be certain agreed terms which are taken for granted and which Descartes called 'simple natures'. These 'simple natures' are words like *unity*, *universality*, *cause*, *motion*, *point*, *equality*, *duration*, *shape*, *magnitude*, *like*, *diverse*, or *existence*. By defining and relating the complexities of a concept to 'simple natures', mathematics could provide the pattern while physics and chemistry, as 'the sciences of number, weight, and measure', could fill it in. By this means, Descartes believed that it would be possible to produce a unitary and definitive system of laws universally applicable throughout nature. Science would thus become a *coda* and not just a catalogue of observed and isolated facts. He turned Man from a believer (i.e. accepting things unchecked) into a skeptic (*de omnibus dubitandum*, 'that all things must be doubted'). But ultimate scepticism is nihilism, a quicksand of doubts; so he had to have a toehold of certainty and it was 'I think therefore I am'.

Man, according to Descartes, exists as the receptacle of *intuitus*. Everything else is an idea which becomes real only because Thinking Man directly apprehends, and proceeds to demonstrate, its existence, e.g. an astral body. But if, by authentication, bodies exist, they must corporately be dependent on an *existent*, which is independent of each of them and of Man, who is the localized point of reference. The *existent* assumed, for Descartes, the prerequisite of God. God, in this sense, is not necessarily the pulpit god of the theologians. Anthropomorphically, he can be the Lawmaker, the Great Mathematician, the Great Architect, etc. Or he can be *nous*.

Nous, as was shown earlier, was Anaxagoras's acknowledgement of the dilemma, 2,100 years before Descartes. It will be recalled that Anaxagoras's elementary particles were the stuff of Chaos, existing in haphazard confusion until they were ordered into Cosmos by *nous*, Reason. Out of Chaos emerged the agglomeration of elementary particles which formed the heavenly bodies, the world itself, the plants, animals, and human beings. These elementary particles provided Man not only with a body and a brain but with the powers to reason. Reason, therefore, was a product of Cosmos but, since Cosmos had to be ordered by Reason, *nous* was a prerequisite.

The dilemma persists even when we accept the 'simple nature' of *eternity* and the *infinitesimal* and assumed that the Cosmos began with the hydrogen atom. How was the microcosmos of the hydrogen nucleus with all its 'manifestations' organized? How did that extranuclear electron, indispensable to all the chemistry of the universe, get into orbit? It is like the old divinity college question 'What was God doing before the Creation?'

Just as there are theologians who claim that Man was made in the image of God, there are those who claim that Man is the Vicar of Reason, a two-legged version of *nous*. That is not very helpful either because we are back with *intuition*, the receptacle of ideas. It raises questions about inspiration, not just among poets but among mathematicians; about imagination, which may be an association of ideas but can often be a spontaneous 'mutation' of experience; about the differentiation of 'brain' and 'mind'; about 'psyche', which some will read as 'soul'. It leads into still-uncharted areas of extrasensory perception. Here are pitfalls for scientists, who, whatever their private opinions about religion or any forms of belief, can, as scientists, only insist upon what they have measured and confirmed.

The reductionist faith

This brings us to the question of 'life'. It would be easier to discuss it if scientists would agree on a definition of 'life'. They do not, but, for the purposes of what follows, we have to make arbitrary assumptions: (1) life is not the consequence of a special dispensation of a deity or a demiurge; (2) to be alive, a thing must not only grow but must be able to reproduce itself; (3) it must metabolize, i.e. extract chemicals from its external environment, select the elements it requires for a self-maintaining biochemical process, and discard what is unwanted.

The exclusion of the deity, theological, Platonic, or Cartesian, is necessary because we are testing the feasibility of 'life' within a measurable system. (And the operative word is 'testing'.) Growth cannot be a criterion because crystals of inanimate matter can grow. A flame metabolizes, i.e. derives the combustible elements from an external source. That inflammable mixture containing carbon monoxide, called water gas, with oxygen, burns to carbon dioxide and water; but we do not, except poetically, ascribe to it the attribute of 'living'.

One recalls that Anaxagoras in his theory of the Unity of Matter did not require two separate assumptions – living and non-living. He gave his 'uniform substances' a built-in life factor. One of the inherent and external properties of all his matter, animate or inanimate, was the organic nature of his infinitesimal entities. With these he was able to construct an infinite variety and a continuous change in the structure of matter. Inert matter, e.g. rocks, consisted of molecules (combinations of uniform substances) in which life was latent. Living plants and organisms consisted of active combinations.

We now know that living things consist of identifiable atoms, no one of which can be characterized by a quality which we would call 'life'. Until 1828 it was thought that organic chemicals were produced only by living matter and that a special 'vital force' was involved. Then Wöhler synthesized urea, an organic substance without any 'vital force', simply by heating ammonium cyanate, a typical inorganic compound. The term 'organic chemistry' still remains a classification of that branch of chemistry which is concerned with chemical compounds identified with substances derivable from plants or animals. It could be described as the chemistry of the compounds of carbon, although, of course, there are substances which contain carbon which have no association with living things.

Since, therefore, there is no 'vital force' and no elements which are inherently 'life-giving', we have to look at living things in terms of their components. That means going back to the atom – not necessarily to the nucleus with its plethora of subatomic 'manifestations' and superlative energies, but certainly to the chemist's atom with the electron orbiting the nucleus.

Although for simplification of explanation in the present instance this 'back-to-the-brickyard' approach is here adopted, it should be pointed out that it is a disputed issue in scientific quarters. The issue was defined by Theodosius Dobzhansky:

> Some scientists cling to the reductionist faith – that the way to know everything is to concentrate on the investigation of the lowermost level, which is consequently styled the 'fundamental' level. A wiser and sounder strategy of scientific research is to gain an understanding of the phenomena and regularities on all integration levels. This is not because some new and irreducible agents manifest themselves in life as compared to the inert matter, or in man as compared to the biological world, as some vitalists wanted us to believe. It is simply because every level of organization shows an integrative patterning of components from the underlying levels, and these patterns are in turn the components of the patterns on the higher, or, if you wish, less fundamental levels. Now, the patterns as well as their components equally deserve attention and study.[1]

In the beginning

Before we can even pretend to climb the disputed Jacob's Ladder from the atom to the sentient being, we have to ask ourselves what chemical 'soup' could have provided the ingredients of 'life' in the beginning.

For this we have no means of direct observation, and never will have. We have Space Machines but no Time Machines. Our space travellers may reach other planets, but they are unlikely (if they are reachable) to be in any way consistent with ours to give us a comparative study of life making. On our own planet even if we found locked away somewhere a pocket of primeval molecules we could never be sure this was a canned version of the original 'soup'.

The Earth, as suggested earlier, is about 4,500,000,000 years old. The first condensations of interstellar gases and dust that formed the Earth and other planets are older – perhaps five to six billion years. About 3,000,000,000 years ago the crust became relatively stable.

There is substantial agreement that the atmosphere was almost wholly lacking in free oxygen, but there were gases such as

1. *The Scientific Endeavour*, Rockefeller University Press, New York, 1965.

methane, water vapour (oxygen locked up with hydrogen), ammonia, hydrogen, carbon dioxide, carbon monoxide, and nitrogen: elemental groupings which could react with each other to form organic molecules. S. L. Miller[2] reproduced, in the laboratory, the circumstances in which gases of the primeval atmosphere might have reacted. He passed electric discharges (60,000 volts) through a gaseous mixture and was able to identify the derivatives as amino acids and other organic molecules. His gases were methane, hydrogen, ammonia, and water vapour. P. H. Abelson[3] found that adding sulphur to similar mixtures led to a variety of organic sulphur compounds. Ultra-violet light would produce even more vigorous reactions than electric discharges, and it has to be remembered that in the beginning there was no free oxygen in the atmosphere. This means that the ultra-violet radiations from the Sun (3,000,000,000 years ago) were much stronger because ozone (O_3) did not exist to modify it.

Most of this chemistry probably took place in the upper atmosphere and the molecules were precipitated by the rains and collected in the seas where they reacted with one another so that the oceans gradually acquired an increasing concentration and diversification of such molecules. They formed aggregates which in turn became more complex. These competed with each other. Some proved more efficient than others in acquiring organic molecules out of the surroundings – like later creatures competing for food – and so grew at the expense of the less efficient. This was a primitive beginning of natural selection.

Meanwhile, changes were taking place in the atmosphere. Hydrogen, the lightest gas, escapes rapidly. By 2,000,000,000 years ago, it would have declined to approximately the present level.[4] In the absence of concentrations of hydrogen, ammonia and methane are unstable and so they disappeared from the atmosphere, being replaced by nitrogen and carbon dioxide. By that time, however, the organic synthesis (as demonstrated in the laboratory) would have taken place and the organic molecules would have been self-sustaining.

2. *Science*, vol. 117, 1953, p. 528.

3. *Science*, vol. 124, 1956, p. 935.

4. H. C. Urey, *Proceedings of the National Academy of Sciences*, vol. 38, 1952, p. 351.

The important thing is that the primitive atmosphere was anaerobic, i.e. without oxygen. The present-day laboratory experiments were carried out in the absence of oxygen. If oxygen had been present at those shotgun weddings (by the electric spark) there would have been no organic molecules; they would have ended simply in combustions. The evolution of cellular metabolism provides internal evidence that much of the early development of organisms must have proceeded in the absence of oxygen.

If, as this argument demands, life first appeared without any help from oxygen, it must have been supported by fermentation, which Pasteur described as 'life without air'. Fermentation makes energy available to a cell by breaking up organic molecules and releasing high-energy phosphates, e.g. adenosine triphosphate, A.T.P. Certain forms of fermentation such as those which produce alcohol have as a by-product carbon dioxide. The release of this gas into the primitive atmosphere by anaerobic forms of life contributed to the evolution of later metabolic processes, including respiration.

The next progression in metabolism was the H.M.P. (hexose monophosphate) cycle. It is essentially an anaerobic process which develops hydrogen from sugar with the aid of energy derived from A.T.P. It also releases carbon dioxide as a by-product. It is important because it is the first example of the metabolic splitting of water. Half the hydrogen in the H.M.P. cycle comes from H_2O. It represents a relatively advanced stage (in the millions-of-years' sense) in primeval chemistry because it is 'getting hydrogen the hard way'. The H.M.P. cycle must indicate an era when practically all the free hydrogen had fled the planet.

Stage Three was probably photophosphorylation – the direct utilization of sunlight to produce A.T.P., with its high energy. This requires pigment, chlorophyll, a magnesium porphyrin to absorb light; and the cytochromes, iron-prophyrin proteins to convert the absorbed external energy, sunlight, into the stored, internal energy, A.T.P.

This chemical evolution (preceding Darwinian evolution of living organisms) is not entirely speculative reconstruction. Our ideas of the origins of species have been reinforced by the discovery of morphological fossils which have imprinted themselves in sedimentary rock formations, but there are molecular relics. They are markers which could be either the end results, or the precursors, of organic systems. There are, for example, carbon compounds with very specific chemical structures which have remained stable through extensive periods of geological time. The strength of carbon–carbon and carbon–hydrogen bonds allows these molecules to survive storage for up to 3,000,000,000 years.

Molecular architecture

An atom consists of a positively charged, small but massive nucleus, in which most of the weight of the atom resides, surrounded by negatively charged electrons, which are much lighter than the nucleus. The nucleus, the 'sun', is about ten thousand times smaller than the diameter of the whole atom, which can be compared to the solar system, with the orbiting electrons as the planets. Since the atom is uncharged, the negatively charged electrons must balance the charge of the positively charged nucleus. Hence the number of electrons must always be equal to the charge of the nucleus expressed in units of electronic charge. This number is characteristic for each kind of atom. Hydrogen, for example, has one electron and one positive-charge unit in the nucleus; helium has two electrons; lithium, three; and so on, up to uranium with ninety-two electrons.

This identification of the number of electrons with recognizable elements give a quantitative definition of quality – the reduction to numbers which scientists love. Bromine, with thirty-five electrons, is recognizably a brownish liquid, with an affinity for other elements with which it forms compounds. Krypton, with thirty-six electrons, is a gas. Rubidium, with thirty-seven, is a metal. Thus a difference of one electron distinguishes a liquid from a gas and a gas from a metal. But how can quantitative differences in structure account for the qualitative differences in properties?

The answer lies in quantum mechanics. When an electron is confined (by the attraction of the nucleus) to a limited region, the wave properties of the electron permit only certain special predetermined states of motion. Each electron is restricted to its prescribed orbit and these orbits determine the configuration of the atom. The orbital patterns, some of them immensely complicated, like the windings of a ball of wool, and their inherent symmetries, determine the behaviour of atoms and their relations one to another. They are the basis of the orderly arrangement of atoms and molecules in crystals. The regularity of a crystal is a pantographic enlargement of the fundamental shapes of the atomic pattern.

The configurations of atoms determine how they fit into each other, like the pieces of a jigsaw puzzle, or the meshing of gears. The pattern of the last electron added or removed decides the configuration and whether the atoms form a liquid, a gas, or a crystal. One electron, added or subtracted, may lead to a complete change of properties.

Precise discussion of the structure of molecules and crystals in terms of interatomic distances began in 1913 when W. H. Bragg and W. L. Bragg used X-rays to determine the arrangement of atoms of sodium and chlorine in the sodium chloride crystal and measured the distance between the centres of the atoms. This was the beginning of X-ray crystallography which has made atoms and molecules visible to the untutored eye in the showcase models constructed from coloured beads or billiard balls, or, at least, they have made the crystallographers' reconstruction visible, because no one has ever seen an atom or an electron.

What crystallography reveals is the pattern produced by the diffraction of X-rays by the wave systems of the particles. When one has learned as much as the analysts have since Von Laue first discovered X-ray diffraction in 1912, one can identify the specific atoms by their wave-frequency characteristics and locate them in relation to each other and in three dimensions. To X-ray diffraction have been added electron diffraction, proton diffraction, and molecular spectroscopy, and the precise structures of thousands of crystals and molecules have been determined.

The carbon atom is particularly accommodating, which explains why carbon compounds are so common on Earth; why they play such a conspicuous role in living matter; and why the chemists have been able to synthesize so many carbon compounds. The reason is that the carbon atom has six electrons, two in close circular orbits round the nucleus and four in elliptical orbits. The four are so arranged that each forms a radial prong away from the centre and the tips of the prongs form a regular tetrahedron. The simplest carbon-based molecule is methane (the main component of cooking gas), in which four atoms of hydrogen link up with the four prongs. The carbon incorporates the single electron of the hydrogen, creating a structure in which the carbon nucleus is at the centre and four protons (hydrogen nuclei) are at the corners of the tetrahedron. Or molecules can be built of two carbons and six hydrogens (ethane); or three carbons and eight hydrogens (propane); or four carbons and ten hydrogens (butane); or eight carbons and eighteen hydrogens (octane), etc. The chains can be extended through an indefinite range of hydrocarbons. The short ones are gases, the longer ones liquids, and the very long ones solids. They are the burning fuels: gas, oil, and candle wax.

Another important group is the carbohydrates. They are like the hydrocarbons but there is an oxygen added at each step in the chain. An accommodating carbohydrate is glucose, which consists of six carbon atoms, twelve hydrogen atoms, and six oxygen atoms.

Another agile molecule-forming atom is nitrogen. It has seven electrons, of which four form a spherical pattern around the nucleus. The other three form symmetrical prongs perpendicular to each other – forward, sideways, upward. When the electrons of three hydrogen atoms mesh with the 'prongs' we have ammonia.

With the obliging atoms, four-pronged carbon and three-pronged nitrogen, we have the two linchpins of another important group of molecules, the amino acids. These are the building blocks of most living matter. They incorporate chemical compounds in considerable complexity, but the simplest, glycine, illustrates an essential characteristic: at one end of the glycine molecule is the amino group, which is nitrogen with two atoms of hydrogen; at the other, carboxyl, which is one carbon atom, two oxygens, and one hydrogen bound together.

All amino acids have this head and tail, the amino group and the carboxyl, which can hitch together, so that amino acids can form long chains. Such chains are called 'proteins', essential to life.

Important in the structuring of molecules is the 'hydrogen bond', discovered by Latimer and Rodebush in 1920. It is a bond formed by a hydrogen atom between two electro-negative atoms. Sometimes the hydrogen atom lies midway between such atoms but usually it is attached more strongly to one than the other. Determinations of the crystal structure of amino acids and simple peptides have shown that molecules of those substances which are closely related to protein attach themselves, one to another, by forming hydrogen bonds between the nitrogen atom and the oxygen atom of the peptide groups. This led to the formulation, by Linus Pauling and his colleagues, of the alpha helix, a configuration of polypeptide chains present in many proteins. This discovery was fraught with consequences which even scientific reticence must call dramatic.

The physical basis of life?

Enough may have been said to show that there is no boundary between physics and chemistry. Molecules are a combination of atoms, and the characteristics of atoms can be quantitatively expressed in precise terms of the particles which comprise them. Ignoring, at this stage, the internal economy of the nucleus, the external relations of the elements, one with another, can be spelled out in terms of the electrons, their energies, their wave frequencies, and their orbits. Sunlight, radio waves, the glowing television screen, the hardest diamond, the most tenuous gas, the burning flame, a lump of sugar, or a nylon thread can all be explained in terms of physical forces.

Can we go all the way and establish the unity between matter and life, reduce everything to atoms and then reconstruct the living processes, right up to the mind which analyses those processes? There are those who say, rightly, 'There is no proof.' And there are those who say, rightly, 'But there is evidence.' The distinction is valid. Isolated pieces of evidence do not constitute proof, no matter how circumstantial. There is no *proof* of a continuum, no Jacob's Ladder from the atom to the sentient being, but we can see some of the rungs of a ladder. Indeed we can start with the rungs of a spiral staircase.

Reference was made to the alpha helix, Pauling's polypeptide. A helix is like a piece of string coiled round a pencil, a configuration which was to acquire a special significance in relation to the structures of nucleic acids: R.N.A. (ribonucleic acid) and D.N.A. (deoxyribonucleic acid).

A brilliant discovery, which has stood the test of time, was made in 1953 – the double-helix structure of D.N.A. James D. Watson, an American biologist, and Francis H. C. Crick, a British physical chemist, working together at Cambridge University, made use of the information then available about the nature of D.N.A., including the X-ray diffraction observations of M. H. F. Wilkins, of King's College, London. They succeeded in constructing a molecular model of D.N.A.

This they found (and innumerable succeeding experiments have confirmed the finding) to be a pair of antiparallel chains wound helically round a common axis, and cross-linked by specific hydrogen bonding. A convenient way of looking at it would be as a spiral staircase with bonds as the steps between.

It is, however, a peculiar kind of staircase. The spiral strands are linked groups of sugar, phosphate, adenine, thymine, cytosine, and guanine. The last four are the critical components which, combined with the others, form what are known as 'nucleotides'. They contain atoms of carbon, nitrogen, hydrogen, oxygen, and phosphorous – prepackaged, as it were. And we have to think of the two strands running in opposite directions – anti-parallel – because the nucleotides on each side are differently orientated.

If we treat the sugar and phosphate as constants, in fixed relationship to the groups, we can identify the nucleotides by their initials – A, T, C, G. Each rung of the staircase, of which there are only four kinds (AT, TA, CG, and GC), is a pair (each pair linked by hydrogen bonds). It makes a difference which nucleotide is on the right side and which is on the left. The nucleotide combinations follow each other in repetitive order, like newspapers running off a press. So you have a template from which a chemical pattern can be repeated.

Here we have to pause and remind ourselves that nucleic acid is a chemical of the nucleus of the living cell. We all of us originate from a single germ cell, the fertilized ovum, about one tenth the diameter of a pinhead. A human germ cell is not very different in appearance from that of a rabbit, a guinea pig or a chimpanzee, or from the germ cell of spinach. But each of these cells carries a vast amount of special information dictating what will grow out of it – a spinach cell does not produce a George Bernard Shaw!

There are specific instructions present in each differentiated cell. How are those directions transmitted from one generation to the next? What is the 'language of heredity'? How are the instructions reprinted and referred, as they must be with each of the cell divisions which intervene between division of the fertilized ovum of one generation and the mating of the next, and by particular specification – 'Be a kidney', 'Be a brain', 'Be a curled hair and not a straight one'? There have to be millions of 'bits' of information, all packed into that minute cell.

More than a century ago Gregor Johann Mendel, studying peas in his Moravian garden, postulated units of heredity. We know them as 'genes'. They are contained in the nucleus of the cell, in 'portmanteaux' called chromosomes. There are forty-six chromosomes in a normal fertilized human egg. The number of genes has been estimated at 'more than ten thousand kinds and less than a million'. Each is present in duplicate except for sex-linked genes which are for the most part present singly in the male. The genes specify the characteristics which differentiate individuals in infinite variety – eye colour, hair form, facial features, etc. They help determine why you are you, and nobody else.

Chromosomes were long known to be made up largely of protein and nucleic acid. Proteins are long-polymer molecules consisting of linear sequences of amino acids, of which there are some twenty kinds. Since the permutations and combinations of twenty amino acids could provide for an abundance of instructions, a reasonable assumption was made by early researchers that the gene was a kind of coded message in protein. Nucleic acid, on the other hand, was regarded as a monotonous kind of polymer (coherent groups of molecules which behave like one molecule) restricted to four nucleotides; this did not seem to provide much opportunity for the innumerable variations which genetic material must produce.

The significance of D.N.A. as the basic stuff of heredity was first recognized through the work of Avery, MacLeod, and McCarty at the Rockefeller Institute for Medical Research in 1944. They were investigating pneumococcal bacteria. It became clear that the type of capsule produced by a pneumococcal strain could be altered by treating it with pure D.N.A. from another strain.

Fresh evidence came during the next nine years from the study of bacteriophages. These viruses are parasitic on bacteria. When their life cycles were studied, it was found that they, the viruses, had a specific heredity. They possessed genes. They consist largely of protein coats containing cores of D.N.A. Electron microscopy suggested that when a virus infected a bacterial cell the protein did not enter the host cell. This was elegantly established by the use of radio-isotopes, the by-products of atomic energy. These are radioactive twins of ordinary elements which, while behaving chemically the same, reveal their presence in compounds by emitting detectable rays. By 'labelling' the sulphur present in the protein and the phosphorus present in the D.N.A. of the virus, investigators were able to show that when it infected the bacterial cell phosphorus was present in the host cell but not sulphur. *Hence the viral genes must be D.N.A., and hence D.N.A. must carry the genetic information.*

Genetic chemistry

Thus genetic information must somehow be determined by the sequence of nucleotides – A, T, C, and G. Since only AT and CG pairings are found in the normal structure, the two chains must carry complementary information. One can think of information being carried as a sequence of nucleotide pairs in the double molecule (hydrogen-bonded) or as nucleotide sequences in each of the strands. Remembering the twenty-six letters of our alphabet which are necessary to form the words of our dictionary, it is difficult to imagine how to get along with only four letters, but think of all the information which has been conveyed by only three 'letters' – the dot, dash, and pause of the Morse code. We have also to consider the amount of D.N.A. in a single human cell. It is estimated that the total D.N.A. in the forty-six chromosomes of a fertilized human egg contains 5,000 million nucleotide pairs. The genetic specifications thus encoded for producing a person from the fertilized ovum would, if spelled out in English, fill 500 volumes of the size of the *Encyclopaedia Britannica*.

We must not let this miniaturization by Nature mystify us unduly. After all, modern techniques can reduce a poster-sized version of the Declaration of Independence to a microdot, no bigger than a printer's full point. Computerized 'memories', using cryogenics, are in principle capable of storing all the information of all the libraries of all the world in a casket no bigger than a cigar box. Cryogenics, in this instance, refers to the behaviour of certain conductors at temperatures near Absolute Zero (roughly, −273° Celsius). Since these conductors offer no resistance to electric currents, the 'wires' can be as fine as you like and the 'valves' can be as small as brain cells. On the reduction principle of the microdot, and reversing, as it were, the electron microscope, it is possiple to etch on a plate five inches by six a printed circuit that mimics the 10,000,000,000 neurons of the human brain and their connexions. By stacking such plates and cross-connecting the parameters in a lattice, one could produce this phenomenal memory capacity. If we find this difficult to grasp, it is because magnification or minification is not an innate sense; it is something to which we have to become rationally accustomed.

A cell, however, is not a piggy bank which hoards bits of information. It passes them on by cell division in which each new cell possesses the same information and organizes itself according to that information; so there must be a system of replication, or reproduction, of the D.N.A. code.

But how? The Watson–Crick model suggested how D.N.A. molecules might reproduce by separation of the double structure into single chains. Since the two original chains are complementary, each carries the information necessary to direct the synthesis of its new partner. The single chain is a template, or jig, which attracts free nucleotides but accepts each in its proper place and in proper order and in no other. It is like an assembly line where Bolt 12 must meet up with Nut 12 and cannot accept Nut 11 because the screw is different.

The forces required to untwist the double helix would not be great, and there seems to be no doubt that it does untwist. One piece of evidence comes from labelling old and new chains, so that they can be distinguished. This has been done with the stable, but heavy, twin of nitrogen. By growing bacteria on a culture medium containing only nitrogen in its heavy form, the D.N.A. molecules, artificially 'doctored' in this way, become 1 per cent heavier than normal ones and can be separated from them by ultra-centrifuge. You grow bacteria in a heavy medium and then transfer them to a normal-nitrogen medium. After one cell generation, i.e. when the total population doubles, the D.N.A. molecules should have, on the Watson–Crick hypothesis, one heavy chain and one light. They have. In the second replication, which doubles the population again, heavy and light chains should separate and each then acquire a light complement. They do.

Another piece of evidence came from Arthur Kornberg and his co-workers in 1956. Using a solution containing magnesium ions and the four nucleotides of D.N.A. as triphosphates and an enzyme and a primer of natural D.N.A., D.N.A. is rapidly synthesized. But single-stranded D.N.A. (obtained by heating ordinary D.N.A.) is twenty times more effective, and the pattern of the D.N.A. is faithfully reproduced. In the Kornberg system, D.N.A. is spontaneously synthesized (without a primer) after a lag of two to four hours, but the spontaneous D.N.A. contains only adenine and thymine (AT) and the other two strands run A-T-A-T-A-T one way and T-A-T-A-T-A the other. If this material is now used as a primer, from a solution which contains all four nucleotides, only AT molecules are produced immediately. The cytosine and guanine (CG) nucleotides are rejected. This is like creating like, and it reinforces, even if it does not prove, the untwined double helix and the template hypothesis.

We have information compiled and stored in the cell. We might even agree as to how that information is copied, but how is it used? If we could take the narrative from there we could boldly entitle this section 'The Secret of Life', but we cannot. The narrative is episodic. It is like picking up a magazine in a dentist's waiting room and getting engrossed in the serial. You find that a critical bit is missing because someone has torn out the cookery recipe on the back of the page. Then you come to the really exciting part and find 'See the next thrilling instalment. Will the D.N.A. messenger get through?' or you search the random pile and find that another exciting development is reported but you cannot find out how it arose because the intervening instalments are missing.

All living systems contain nucleic acid (D.N.A. in most cases) and protein. The proteins are of many kinds compiled from the permutations and combinations of twenty amino acids and serving many functions, structural and chemical. The key proteins are enzymes, organic catalysts that speed up and regulate (without themselves being changed) chemical processes on which life depends. It seems that the chief way in which genes determine the character of the organism is by ordaining the synthesis of specific enzymes, along with other proteins. Each gene is in fact a recipe for a protein in the coded chemical language of D.N.A. How does the recipe become converted in the cell's protein factories?

Let us consider a working hypothesis as to the process of protein synthesis. For each protein capable of being formed, there is in the nucleus a specific segment of D.N.A. that carries the information as to how the amino-acid subunits are to be assembled. That segment is analogous to the punched tape of computer programming. Let us agree to call that code segment a gene.

Then take the gene for human haemoglobin. Each molecule of this oxygen-carrying red protein, indispensable to vertebrate life, consists of four chains of protein and an equal number of haem groups, each containing iron. The protein chains are of two kinds – alpha and beta. There are two of each, the members of a pair being identical and each made up of about 150 amino-acid units of different combinations but arranged in a precisely determined sequence. There is reason for believing that for each of the two chains a gene exists consisting of something of the order of 500 D.N.A. nucleotide pairs.

From the gene in the cell nucleus, the information is transferred to a 'messenger' which has been formed by D.N.A. directing four kinds of ribose nucleotides to create R.N.A. R.N.A. is 'ribonucleic acid', a single-stranded helix. This messenger form of R.N.A. moves from the nucleus to the cytoplasm of the red blood cell. There it gets incorporated in submicroscopic bodies made of structural protein and structural R.N.A. The information which has now arrived is immediately put into effect in ordering the amino acids into the proper sequence to make haemoglobin protein chains, through peptide linkages. Coming off this production line, the alpha associates with the beta and with the haems and goes on to function as an oxygen-carrying red protein.

Sometimes that protein departs from the norm. Indeed, there are at least a dozen modifications of human haemoglobin known, many of them established as genetic deviations – a different D.N.A code. These deviations, or mutations, once they occur, go on repeating themselves in cells and in generations of beings. It is like a misprint which goes through a whole edition of a newspaper.

Mutation is believed to be due to the alteration of the nucleotide sequence in the D.N.A. structure. Once the sequence has slipped, the change goes on repeating itself as a new version of the code. This suggestion is plausible because it is consistent with what is observable when mutations are artificially induced. This can be done by bombardment of living material with high-energy radiations. In 1927 H. J. Muller demonstrated the genetic effects of X-rays; and radiation hazards to future generations have been headline news ever since the atom bomb went off. One can consider those high-energy radiations as *force majeure*, jerking the nucleotides out of position and causing regroupings. Moreover, many chemical substances are capable of changing the genes. The mechanism can be grasped by considering the action of nitrous acid. This specifically oxidizes amino groups. In this way it can change a nucleotide in D.N.A. in such a way that its hydrogen bonding is modified and different conjunctions can take place.

Natural mutations – interference with a continuing code, or genetic traits – are a mechanism of evolution. Positive evolutionary progress depends on favourable mutations. If the alteration in the code becomes a 'printer's pie' (xvz/4ruf'½ 6wxxxx), it will repeat itself, but it 'won't work' and the cell function will cease. If it is a plausible but 'wrong' grouping, it will be genetically deleterious. Nature rejects monstrosities and abnormalities, but, also biologically unfavourable mutations ('wrong' groupings) mean that organisms have reduced reproductive fitness. The errors in D.N.A. may not be corrected in replication, but the lines of descent carrying the flaws are bred out of existence. How long it takes depends on the degree of reduction in reproductive fitness. This we call 'natural selection'.

Favourable mutations are those which make improved use of the environment whether that environment is in the body or in the external world from which the elements to be metabolized must be derived. Reproductive fitness, therefore, implies new strains which will gradually replace the ancestral types from which they started.

14 Proton to Paragon?

The ascent of life

The characteristic of a living system, according to John H. Northrop, is 'to use energy to carry out the synthesis of more of itself.'[1] This rules out autocatalytic reaction (in which a chemical which facilitated the reaction of other chemicals emerges as a product of the reaction). It also rules out crystals, because while they grow by reproducing themselves, they do so by giving away energy. That is to say that, on the energy gradient, they run downhill while the living organism climbs uphill. The analogy in hydraulic terms would be between a waterfall which drops water to release energy and pumped storage which lifts water uphill to provide a head of energy. In energy exchange, therefore, we can speak of the 'ascent of life'.

While there may be philosophic difficulties in defining what is meant by 'life', there are certain things which everyone would acknowledge as 'living': plants, animals, humans. Plants supply the sustenance of animals and humans by fabricating carbohydrates and proteins in forms which creatures can digest and convert to the needs of their own metabolism.

The plant is a factory which derives its energy directly from the Sun and its raw materials from the atmosphere, from water, and from the chemicals of the soil. This involves photosynthesis which is largely an integration of developments already described – the H.M.P. cycle, including its metabolic splitting of water and photophosphorylation, the absorption of light by chlorophyll, and the conversion of the absorbed energy to A.T.P., which, in turn, uses it to bind molecules together in increasing complexity.

Chlorophyll is the green pigment which is present in plants and algae. Richard Willstätter won the Nobel Prize for Chemistry in 1915 for showing that it consisted of four components, two green

1. *Annual Review of Biochemistry*, vol. 30, 1961.

ones (chlorophylls *a* and *b*) and two yellow ones (carotene and xanthophyll). The base of the chlorophylls is a porphyrin ring, and porphyrin synthesis starts with glycine and acetic-acid compounds which can readily be assumed to have been present in the primeval 'soup'. Very early, therefore, molecules could have been formed of porphyrin linked with magnesium, the starting point of chlorophyll, and of porphyrin linked with iron, which forms the red pigment of the blood.

The plant pigments have the property of trapping the electromagnetic waves of sunlight and of transferring this imported energy to the chemical economy of the living process. About 2 per cent of the sunlight falling on a green plant is transformed into chemical energy. The rest is lost by reflection or by absorption by other pigments which do not put it to work. In the laboratories, the efficiency of photosynthesis can be as high as 40 per cent.

Photosynthesis can be defined as converting carbon dioxide into carbohydrate – sugar or starch. The first stage is to use the incoming energy to split water into hydrogen and oxygen. The molecule of oxygen thus released becomes the atmospheric oxygen which we breathe. This free oxygen was not present in the original atmosphere, so that our human respiratory processes depend entirely on what the photosynthetic action of plants has produced in the interval. The second stage is that in which carbon dioxide plus four atoms of hydrogen become carbohydrate plus water.

The second stage is now completely comprehended as a result of discoveries of Nobel prizewinner Melvin Calvin. Using radioisotopes, he produced a blueprint of the process as detailed as the flow sheet of a chemical industrial plant. Carbon-14, which is radioactive, was added to the carbon dioxide which the plants were using. This labelled carbon, identifying itself by its rays, could be traced through each of the progressive chemical combinations which went on in the plant. At set intervals, the growth process was arrested (i.e. enzymic action inhibited) by the plant being treated with boiling ethanol. Extracts were made, at each interruption, and the constituent compounds analysed by chromatography and radioautography.

Chromatography – a term, incidentally, which was invented in the early days of separating the chlorophyll pigments – employs the method of putting chemical materials to be examined in a solvent and allowing them to suffuse through an absorbent material (like a blot of ink on blotting paper). Since different molecules move at different rates, they sort themselves out at different spots. In this case – two-dimensional paper chromatography – they were revealed like plots on a chart, each recognizable as a different fraction in the chemical process. Placed on a photographic plate, the presence of Carbon-14 revealed itself as a ray shadow. On the analogy of the flow sheet of an industrial chemical process, isolating and examining the various molecules is like sampling each stage of a fractionation process.

As a result Calvin and others were able to show, and later workers were able to reproduce and confirm, the complete path of carbohydrate production in photosynthesis, with all intermediate and enzymatic reactions from carbon dioxide to sucrose and starch, with a clear indication of the route to the amino acids and the proteins beyond.

One pathway leads to the Krebs Cycle (Sir Hans Adolf Krebs, Nobel Prize, 1953). This is sometimes called 'the citric-acid cycle' and defines the series of chemical reactions by which citric acid is oxidized in living organisms. The combustion of foodstuffs, proteins, carbohydrates, and fats provides energy (or releases energy, if we accept that it has originally been supplied by the Sun and stored in the molecules of those substances). The earlier stages of the oxidation process lead to the formation of acetic acid. In a further stage, the acetic acid is converted to carbon dioxide and water – back to first base! The reactions of the cycle, in combination with others, help to form the carbon structure of glutamic acid, an amino acid, which is the parent substance of other amino acids. These, marshalled in many ways, are the groups which build proteins.

All amino acids isolated from protein molecules belong to a single family of compounds. At the centre of each group is a carbon atom. Hitched to it, on one side, is a three-atom combination: one nitrogen and two hydrogens. (This is the 'amine' group, so-called because it is related to ammonia.) On the other side is a combination of one carbon, two oxygens, and one hydrogen atom. On the third side is a single hydrogen atom. On the fourth side is a side chain, which may be quite complicated, with as many as eighteen atoms. This side chain is what differentiates the amino acids, of which twenty-one are fairly common, though others have been identified. Amino acids combine to form a protein by having the amine of one connected to the carboxylic-acid group of its neighbour. So an amino-acid chain is formed with the side chains, like feelers reaching out to attract other elements or molecules. Those side chains make each type of protein molecule different from all others. In the living system protein molecules are broken up to be reassembled. For example, the proteins of a steak are split by the gastric juices of the stomach and are redistributed as smaller chains of amino acids called 'peptides'.

The molecule of insulin, so literally vital in the treatment of diabetes, is a polypeptide made up of about fifty amino-acid groups. By the late 1940s it had been shown that insulin had a molecular weight of approximately 5,500. That means that the molecule weighs 5,500 times the weight of an atom of hydrogen. It was found to consist of two chains of amino acids, zip-fastened together by the amino acid, cystine. Chain A consisted of twenty-one amino acids and the other, Chain B, consisted of thirty. This was established by breaking down each chain separately and isolating the different aminos by paper chromatography. This was like spreading a thirty-letter alphabet on a table in a word game. It did not give any clue as to how the 'letters' were formed into 'words' or 'words' into 'sentences'. The twenty-one amino acids in Chain A might be arranged in 2,800 million, million different ways. The thirty amino acids in Chain B offered 510 million, million, million, million alternative arrangements.

A group, led by Frederick Sanger, the British biochemist, tackled the 'word-making' to determine the exact arrangements in a molecule of insulin from ox pancreas. They did it by partially breaking down each chain into fragments containing two, three, or four amino acids, by use of appropriate acids and enzymes (after the fashion of the gastric juices). They were now dealing not with letters but 'syllables' and 'words', groupings which chemically could combine only in a limited way. It took eight years to unscramble and reassemble the amino acids in their proper order. By 1955 they had put the fragments together and had determined the structure of the intact molecule. This was the first time that the structure of a naturally occurring protein had been exactly defined. For this, Sanger was awarded the Nobel Prize for 1958.

If we consider the simplest of all living units, the bacterial cell, it consists of 5,000 different kinds of proteins. Each has a special function. Some are fibrous and are knitted together to form the membrane which encloses the cell and the internal partitions. Others are long chains, coiled up like a skein of wool. These are globular proteins and they make up most of the jelly-like content of the cell. These proteins within the cell are chemically active. Some act as enzymes, biochemical catalysts, which decompose nutrient molecules and rearrange their elements into amino acids and nucleotides.

The nutrient molecules are imported into the cell. A bacterial cell will grow and ultimately divide into two new cells if it is immersed in a solution of sugar, phosphate, and ammonia. Sugar and ammonia are simple molecules and yet the internal mechanism of the cell can contrive from them twenty kinds of amino acids and four nucleotides and assemble them in correct order to become 5,000 proteins and the nucleic acids necessary for replication.

Energy is a first requirement. When amino acids are formed and marshalled into proteins, the groups have to be energetically moved around and joined together.

Sugar contains energy. Certain specific proteins in the cell attract sugar molecules to their surface and cause them to disintegrate into groups of atoms which are reshuffled to form carbon dioxide and water. This is what happens if we burn sugar and release energy in the form of heat. In the cell the heat is captured by molecules of adenosine phosphate. When energy is acquired the molecule attaches another group and becomes adenosine triphosphate – A.T.P. When energy is needed for some other chemical process within the cell, the A.T.P. (in high-energy quantum state) delivers it and becomes adenosine diphosphate –

A.D.P. (low-energy quantum state). This charging and discharging of energy can go on as long as the carbohydrate source is made available. In the plant, as we have seen, the carbohydrate is synthesized by energy from the Sun. In the bacteria (or the human, for that matter) the molecules are derived, prefabricated, from a nutrient.

Energy, however, is not the only thing which has to be imported from an environment external to the cell, whether in the plant, the bacterium, or the higher organism. The mechanism of the cell can only work if the molecular raw materials are available. Therefore, all the atoms of all the amino acids, of all the proteins and of all the nucleotides must be purveyed like the ingredients of a prescription.

What we call 'deficiency diseases' are due to the absence of essential molecules, e.g. vitamins, and sometimes of infinitesimal trace elements. A cell will use molecules most easily assimilable. If the surrounding medium becomes depleted of such staples, the cell will cease to grow or to divide, and that species of cell will become extinct. The cell species which survive are those which, somehow, can adapt to changing conditions. The chemical requirements for survival remain the same, but they have to be got the hard way. N. H. Horowitz[2] has suggested that the biosynthetic sequences evolved by adding one enzyme at a time.

An organism running short of some essential nutrient hitherto readily available might develop an enzyme for performing the last step in the synthesis of the precarious substance so that the organism's needs would then be transferred to the immediate precursor. As that in turn became depleted, the organism might evolve an enzyme for the previous step in the synthesis and so on until the whole synthesis could be performed from simple and readily available precursors. A gross analogy would be: if you cannot buy bread, buy the flour to bake it; if you cannot buy flour, buy grain and mill it, etc. – not the easiest way to get food but a way to an increasing self-sufficiency. In the case of the cell, each enzymic refinement offered a better chance of survival.

2. *Proceedings of the National Academy of Sciences*, vol. 31, 1945, p. 153.

The question mark – the virus

Molecule by molecule, cell by cell, we seem, thus far, to be establishing the characteristics which distinguish the animate from the inanimate, but now we run into the anomaly which is the virus. Is the virus a living agent?

Around the beginning of the twentieth century viruses were discovered – first the plant virus of tobacco-mosaic disease; then the virus of foot-and-mouth disease in cattle; then the yellow-fever virus, affecting humans. They were so small that they were beyond the range of the most powerful microscopes then existing and passed through the finest filters. They grew and reproduced within specific living cells and they were capable of mutation. There seemed to be no question that viruses were the smallest living organisms. Around 1930, the sizes of different viruses were determined with some precision. Some were actually smaller than certain protein molecules. Then in 1935, Wendell M. Stanley isolated the tobacco-mosaic virus in the form of a material which could be crystallized. It was a nucleoprotein – nucleic acid with a simple protein. As a crystal, the virus was inert; it could not reproduce itself; it was as inanimate as a piece of quartz. Yet this inert substance in contact with a living host was capable of producing a characteristic disease. This it did by reproducing more of its like.

A virus is a sac 'woven' of simple proteins and filled with nucleic acid. It has not got the complicated mechanisms which produce amino acids or A.T.P. or nucleotides. It attaches itself to the outer membrane of the host cell. It diffuses its nucleic acid into the cell and just uses the cell's amino acids and the nucleotides and the energy of the A.T.P. to reproduce itself. It borrows enzymes from the host cell. It borrows such nucleotides as it needs for its nucleic acid and such proteins as it needs for mass-produced coats. It multiplies with such speed that a new horde of viruses bursts the membrane of the host cell, killing it, and erupting to attack other cells in destructive disease. Polio virus was the first human virus to be obtained in crystalline form.

Regardless of certain mental restrictions that may differ from person to person [Wendell M. Stanley has said], I think there is no escape from the acceptance ultimately of viruses, including crystallizable viral nucleoprotein molecules, as living agents. This must be done because of their ability to reproduce or to bring about their own replication. Certainly the essence of life is the ability to reproduce, to create a specific order out of disorder by repetitive formation with time and of a specific predetermined pattern and this the viral nucleoprotein molecules can do.[3]

Many viruses contain D.N.A., which the modern school of molecular biologists claim is the code by which that predetermined pattern is passed on. The chain in the virus is short; the 'spiral staircase' has only 100,000 steps, but if we accept the claim, that is all the 'information' the virus needs to start a deadly epidemic.

Molecules incorporated

The percipient reader will have noticed that we have done the molecular chemistry without providing the laboratory bench, the flasks, and the retorts. Those who question the suggestion that all life can be traced back to fundamental particles will point out that the organization of the flasks and the retorts is more difficult to account for than the reactions which take place in them.

Atoms can combine to form molecules only in proximity close enough for the orbital electrons to interlock and the energy quanta to jog other electrons into chemical activity and produce molecules. This cannot happen casually in expanse, either of the atmosphere or the oceans. In origin, there had to be a laboratory bench and later a complicated system of flasks and retorts: cells, all differentiated but, in the organism, all interdependent.

One theory suggests what the primeval 'laboratory bench' may have been. J. D. Bernal has speculated that clays formed by the precipitation of solutions of alumina and silica provided a likely base for the fixation of chemicals in the early stages of the formation of the crust of the Earth. The mud of the sea bed would absorb organic substances on its surface from the water medium. More than just providing propinquity, the clays would act as a filter, like a chromatographic column through which dissolved

3. *The Origin of Life on Earth*, ed. A. I. Oparin and others, Pergamon Press, New York, 1959.

materials would pass and from which molecules would be separated into different zones, where those with special affinities would react. Moreover the beds would be continually overlaid by new layers. On this hypothesis mud becomes an apparatus in which biosynthesis could occur.[4]

Scientists use the terms *in vivo*, meaning 'in life', and *in vitro*, meaning 'in glass'. The first refers to the observable processes occurring in the living plant or animal; the second, to analysing, measuring, and experimentally reconstructing such processes outside the living plant or animal. *In vitro* explicitly says that there is a glass vessel – a test-tube, a flask, or a dish – within the confined space of which reaction and synthesis can be controlled; *in vivo* that confined space is a cell the diameter of which is reckoned in microns, or thousandths of a millimetre. In this microminiature apparatus the complex chemical processes are elaborated. Life, self-sustained and self-reproducible, exists at cell level. Many living forms do not go beyond this single-capsule stage and maintain an autonomy independent of tissues, organ, or organism. Similarly, many cells, removed from interdependency in an organic community, will revert to an independent state under culture *in vitro*.

The human cell, which eventually 'adds up' to the eye which is reading this and the brain which is trying to understand this, is extremely complex. At its centre is the nucleus, separated from the cytoplasm of the cell by a porous membrane. The nucleus contains a nucleolus, a small dense body of nucleoprotein. This nucleoprotein (R.N.A.) directs protein synthesis through 'messages' received from the chromosomes, or rather from the genes which are the 'beads' on the chromosome string. Nucleolar R.N.A. 'communicates' through the membrane of the nucleus into the cytoplasm, which is both a 'factory' and a 'warehouse'. The 'factory' is the endoplasmic reticulum adjoining the nucleus. It consists of labyrinthine spaces, along the membranes of which R.N.A.-rich particles, the ribosomes, proceed to make protein. The 'warehouse' consists of inclusions in the cytoplasm which have the elements necessary for making the protein and nucleic acid and which store the cell products prior to cell division. The

4. Oparin, op. cit.

cell's power batteries providing the energy for the processes are the mitochondria which obtain the nutrients and oxygen necessary from outside the cell. When the processes are sufficiently advanced the molecules begin to aggregate as clumps which form into filaments: the chromosomes which will be shared by the daughter cells. At this stage, division takes place. There are now two sets of everything which was in the original cell, arranged in compartments so that when the membrane of the cell elongates owing to the increasing pressure within, it divides at the waist and self-seals into two packages instead of one. (This is the 'fission' which the physicists borrowed from the biologists as a term to describe the splitting of the uranium atom following invasion of the atomic nucleus by a neutron.)

All this has been systematically learned by the scientists. Even before 1838 when Schleiden and Schwann compared notes and found the similarity in structure of plant cells and nerve cells, the cellular system had been noted by microscopists. What the observers could see through the lens and painstakingly draw was later transformed by histochemistry. In the last half of the nineteenth century techniques were evolved for fixing and staining cells. Fixation means killing and hardening a cell. Staining came from the dyestuff industry. It was found that cell structures could be dyed, just like the fibres of cotton, wool, or silk, so that features could be seen in bright relief, in glorious Technicolor. It was also found possible to identify specific chemical substances within the cell by use of colouring agents; e.g. the recognition of starch by its iodine reaction. Researchers also made elusive structures visible by impregnating them with opaque materials – the ultrafine structure of the cell could be identified by coating it with a delicate film of gold or silver.

The mystery of the membrane

Recent advances in the analysis of cellular organization have been achieved by two technical developments: electron microscopy and cell-fractionation procedures. The first means an improvement, four hundredfold, over the resolving power of light-wave microscopy, and the possibility of carrying structural analysis down to the molecular level. The second allows the isolation in manageable quantities of most subcellular structures for their chemical and functional analysis *in vitro*.

It is, therefore, possible to study the cell, surrounded by membrane; the mitochondria (the energy transfer system), surrounded by membrane and, within it, two chambers divided by membrane; the endoplasmic reticulum, surrounded by, and internally divided by, membrane; and the nucleus, surrounded by, and internally divided by, membrane. The recurring word is 'membrane'.

One might glibly say that membrane is to *in vivo* what glass is to *in vitro*, but that would be wholly inadequate. It is, like a glass vessel, the enclosure and confinement of the various processes to that essential propinquity without which molecular associations and reactions would not occur, but what happens if the 'vessel' takes an active and indispensable part in the reactions?

Fixing, staining, etc., by which much was learned about structure, meant killing and denaturing the cell. The membrane was recognizable as a partition or as a vessel from which the chemicals could be extracted. It was even possible to examine it like the skin of a sausage from which the meat has been removed.

Much of the information so obtained about membranes was bound to be misleading or overlooked. It was like Edgar Allan Poe's 'Purloined Letter' – it was so conspicuous, no one noticed it; it was asking to be read, but the incurious saw it only as an envelope. Scientists have now become more inquisitive.

The outer membrane of the cell is a stratified structure consisting of a layer, only two molecules thick, of lipids, sandwiched between protein films – like a raincoat rubberized between the outer fabric and the lining. It is $80 \times \frac{1}{10{,}000{,}000}$ of a millimetre thick. A lipid is insoluble in water and would impose an impermeable barrier to exchanges between the aqueous contents of the cell and the aqueous surroundings of the cell which the texture of the fibrous protein might allow. Yet demonstrably the cell imbibes and excretes and breathes through the membrane. This selective permeability can be explained as a mosaic of functional units – of pores, valves, and energy-driven pumps – each dealing with the transfer of individual molecules or ions. Within the cell the membranes are even more tenuous and not only provide partitions to contain specific processes but chemically participate in the processes they are containing. In other words (on the *in vitro* analogy) the walls of the cell 'flask' are not inert, like glass, but have an active role.

This is the chicken-or-the-egg dilemma (which came first?) expressed in millicron, millionths of an inch, dimensions; the processes depend on the membrane but the membrane depends upon the processes, i.e. the protein of the membrane derives from amino acids, the assembly of which requires the 'flask' which the membrane provides.

In speculating how the membranes might have originated, it is tempting to use the macroscopic examples of the raindrop and the soap bubble. The water vapour in a super-cooled cloud needs a nucleus on which to condense. (In artificial precipitation, clouds have been 'seeded' for this purpose by dispersing silver-iodide molecules; carbon dioxide 'dry ice'; and even common salt.) With the toehold of such a nucleus, the H_2O molecules combine and the outer molecules (interface) form a self-sealing film which encloses the liquid. The soap bubble is a balloon in which the internal pressure stretches the soap molecules into a film in a state of surface tension. A surface, thus presented, would provide the base on which appropriate molecules (like soot on a raindrop) could agglomerate and combine to form a sheath.

The requirement of this sheath, or cell coat, would be to let simple chemicals through but to keep chain molecules inside, because, as we have seen, the business of replication and differentiation of the living processes depends on those chains, whether they are of nucleic acids, polypeptides, or proteins.

Chemical evaluation

We have continually to remind ourselves that our time scale is measured in billions of years, which allows ample opportunities for chance happenings, with countless billions of discarded 'experiments', of combinations which did not persist. Atoms met at random in the atmosphere or in the sea. They could either attract or repel each other. If they had a chemical affinity, i.e. if their electron orbits could intermesh, they would form molecules.

Some of those molecules would by their configuration incorporate other elements or federate with other molecular groups. Among those chance compounds were some which were to play their parts in living processes – sugar, nucleotides, and amino acids. Carbohydrates were formed by sunlight in much larger quantities than amino acids or nucleotides. The formations were slow and they were continuously disrupted by the intense ultra-violet rays, bombarding a planet still unprotected by ozone. In the seas, however, at layers beyond the impact of ultra-violet rays, they began to accumulate. In favourable circumstances – in pools or, as Bernal imaginatively suggested, in the natural chromatographic columns of the clay beds – they would be in close enough conjunction to form chains. Those could have been short chains of nucleic acid or proteins, but as isolated, nonreproductive combinations.

Some time, in those billions of years, a combination of favourable circumstances found a special kind of nucleic-acid chain in a medium of proteins which made nucleotides, not in a haphazard fashion, but in an orderly arrangement which would encourage what we would now call life proteins. In sorting out the nucleotides the chain would extend itself to provide the pattern of other proteins which, in series, would provide the prescription of 'life'. Five types of protein would be necessary: a protein capable of making nucleotides; a protein producing amino acids; a protein which would 'burn' sugar so that its energy could be stored and transferred, through A.T.P.; a protein which would form special molecules (chlorophyll) to synthesize sugar with the help of the Sun's energy; and a protein which would form the membrane to capsulize the system and intensify it.

The first would facilitate nucleic-acid formation; the second would increase the amino-acid supply which previously had been dependent on the slow and dispersed method of ultra-violet production; the third, by providing energy carriers, would weld one molecule to another in chains; the fourth would make the system less dependent on the random availability of sugar in the medium; the fifth we have already discussed in terms of the membrane.

Given those components, we have something as simple as a

bacterial cell. In a satisfactory medium, e.g. a puddle with sugar and simple chemicals, it will 'live'. That is to say, amino acids and nucleotides are formed within the unit; the amino acids are 'jigged' into place by the nucleic-acid template to form specific proteins, and the nucleotides are used by the nucleic acid to form replicas of itself until the reproduction of proteins and nucleic-acid chains becomes too much for the confines of the cell, which bursts or divides; the fabricated units, now independent, repeat the process.

If this were all, and an original cell had merely reproduced itself *ad lib*, 'life' might have been just a cellular slime. Obviously, if one type of unit could 'happen' in this way, so could others with differences. Another factor, however, produced differentiation. This was the mutation of the structure of nucleic acid. Mutation can happen in at least two ways. Errors occur in the process of replication. If the new template is different it may not be able to produce the necessary proteins, and the unit in which the change has occurred will just cease to function. If, however, the modified nucleic acid can produce proteins in spite of the change, the altered prescription will be repeated in the cell progeny. It will be a new kind of cell which may be viable in a different way. Another form of mutation can occur when the nucleic acid acquires additional nucleotides and adds them to the chain. These additions repeat themselves in the replicas. The additional 'prescription' may produce additional new proteins, and these may improve the efficiency of the cell. For example, they may make better use of raw materials which are in short supply, and they will thus multiply faster than the previous type of unit which will presently die out because it will lose in the competition for nutrients. This, at cellular level, is the survival of the fittest.

Most changes are towards higher differentiation, through longer chains of nucleic acid which produce proteins for more and more specialized functions. This means a system of 'sub-contracting'. The functions are divided among different cells and the viable unit becomes multicellular. In a system of interdependence some plant cells build up the framework – the stalk; some can collect nutrients from water or from soil – the roots; some trap the energy of the Sun – the leaves. The 'instructions' for all these, however, are in the germ cell of the seed and are allocated by the genes, which, as has been discussed, are segments of nucleic acid.

Thus far we have been discussing replication merely as the multiplication and division of chemical molecules inherent in an individual cell and of mutations as accidental changes in the original 'blueprint' of the cell. A much more efficient way of development occurred when two units united before replication and made use of a mixture of their nucleic acids. The template then combined the assimilated 'instructions' of both units. this system of sexual replication had the great advantage of combining new and successful trends, of pooling the results of mutations which had already occurred in ancestral cells, and thus accelerating better adaptation.

Plant life was an essential precursor of animal life, not only as a source of food but in supplying the oxygen which was lacking in the primeval atmosphere and which was supplied as follows: the building up of nucleic acid and protein needs energy. In the absence of atmospheric oxygen, the source of energy was in the sugar molecule from which it had to be released by fermentation, an inefficient way of getting energy. As has been mentioned, as plants developed the mechanisms of photosynthesis, they were able to split water into hydrogen and oxygen and they released molecular oxygen into the atmosphere. When free oxygen became available it was much easier to burn sugar in the cell and to store the energy in the A.T.P. molecules. New units originated which made use of atmospheric oxygen. Cell respiration meant that, with additional energy available, molecules could be formed more readily and new cells could grow faster.

New and differentiated cells meant, in combination, new and differentiated tissues, which in turn meant new and differentiated organs, building up to new and differentiated organisms.

Biological engineering

Evolutionary processes became biological engineering. Specific bone cells provided the structural units for the skeleton. Cells formed tissues which could contract and expand and could employ the energy which respiratory oxygen made available and muscles resulted. In large multicellular units the oxygen of the air could not readily enter the body cells. Oxygen had to be piped to them from cells which could absorb oxygen: the lung tissues. There developed a system of arteries in which a liquid could flow to the various parts carrying red blood cells which absorbed, readily, the oxygen inhaled by the lungs, and acted as porters. Thus the multicellular organisms we call animals could make much better use of energy in the molecular chemistry of the cells, so that cells proliferated and variegated in a faster form of evolution.

The greatest step forward was the development of the nervous system. This was a special combination of interlocking cells capable of transmitting stimuli – quanta of cell-generated electricity – from one part of the multicellular unit to the other. Special cells, sense organs, through connecting neurons, could be made to influence muscle responses. In the multicellular unit, unrooted and mobile, the nerve senses could direct the muscles to react to environmental changes and thus coordinate locomotion with light or sound, feel or smell, temperature or taste. The organism could move towards light; recognize food by its smell or shape; avoid danger; or seek warmth or coolness.

The development of a nervous system for better coping with the environment was so effective that any mutation or sexual combination of characteristics increasing the range and scope of the communications network gave the benefited species better chances of survival. Whereas, before the development of the nervous system, the living unit's reactions to the outer environment had been entirely dependent on chemical structure – a contiguous molecule of the right shape – there now developed 'behaviour'. At first this was just reaction to stimuli but the nerve cells began to build up into an organ which could store the effects of the stimuli. This storage is what we call 'memory' and, if it is transmitted to the next generation, we call it 'instinct'.

Memory and learning mechanisms need not be very complicated. Modern computers and feedback systems can remember past situations or instructions and determine later actions on the basis of 'experience'. A machine with a few thousand transistors can perform remarkable feats of memory, remembering situations and avoiding them. The brain of an insect can possess a hundred thousand brain cells (the equivalent of the transistors) and the human brain consists of over 10,000,000,000 cells.

This agglomeration of cells, convoluted in the human brain in a casket, the skull, is what characterizes *Homo sapiens*, Thinking Man. He has his sense organs of sight, hearing, taste, smell, and feeling through which he maintains contact with the external world. He has memory, his stored awareness of sensory experience, but he has a great deal more. He has reason; he has imagination and he has emotions in which his nervous system interacts with his body chemistry, his glandular secretions.

All these add up to a process called 'thought', which distinguishes him from all other creatures. By wise thought, he can become increasingly master of his environment.

Cogito ergo sum – 'I think therefore I am'. But he also thinks about where he is and what he is, about his spatial relationship to the Cosmos; about his planet; about atoms and fundamental particles, and about the molecular chemistry that adds up eventually to the brain cells which enable him to think.

It is inconceivable that any scientist today would accept the Adam-and-Eve hypothesis, or regard man as a special creation. A confident reductionist would probably follow George Beadle, who accepted that in the beginning there was a universe of hydrogen and said,

> Thus it is clear that there is a natural sequence: hydrogen; helium; beryllium-8; carbon; oxygen; other elements; water; other inorganic molecules; simple organic molecules; more complex organic molecules, like nucleotides, amino acids and small proteins; nucleic acids capable of replication; nucleic acids protected by protein coats; viruslike systems with protein coats serving catalytic functions; multigenic but subcellular plants and animals; and, in our line of descent, Man himself. The sequence could have arisen by steps no one of which need have been larger than the individual mutational steps we know in today's living systems.[5]

There you have a Jacob's (or Beadle's) Ladder from the proton

5. *Communication in the Modern World*, Granada Guildhall Lectures, British Association, University of London Press, 1961, p. 38.

to the paragon, but for those who find the ascent too steep there is that reminder (already quoted) by Theodosius Dobzhansky:

every level of organization shows an integrative patterning of components from the underlying levels, and these patterns are in turn the components of the patterns on the higher, or, if you wish, less fundamental levels. Now, the patterns as well as their components equally deserve attention and study.

That, of course, is the description of the organizational structure of the life sciences themselves. There is not a straight ladder but a series of plateaux, and on each plateau scientists are intensively studying the patterns, from the crystallographers assembling the atoms in D.N.A. in the basement, to the behaviourists surveying mankind from the penthouse. They are, naturally, interested in what is going on down below and up above, but they have preoccupations of their own. They have more than enough to keep them busy in their own chosen field, whether it is the eccentric behaviour of the cell which produces cancer; or the action of thalidomide on the genes which produces offspring without limbs but otherwise healthy; or turbulence of brain cells which we call 'insanity'; or any one of the hundreds of problems within the 'pattern' they are observing, or the specific phenomenon which they are trying to reduce to measurements.

The nature of Man

Great arguments are generated in debates about whether scientists will ever produce 'life' in a test tube. They are arguments to little purpose because even if a 'living' unit were synthesized, even if it were a multicellular Frankenstein creation, someone will say 'But it is not Einstein' or 'It is not Beethoven'. Supposing that the scientists had synthesized all the ingredients, the nucleic acid, the proteins, the enzymes, and even the membranes and had manufactured a cell? Supposing, by an even greater advance, two cells were married and went on replicating, with all the genetic information being passed on, to form specific cells, specific tissues, and specific organs, including the brain, and man has emerged by total synthesis?

There remains the mystery which lies between 'brain' and 'mind', between the 'information' received and stored (as in a computer) and the conscious capacity not only to exercise logic – relating number to number and fact to fact – but to have imaginative insight. Imagination is not just memorizing and building up a picture of something observed by a self-generated insight, a conceptual idea. A machine can assemble facts in an infallible memory, but it has not got that quality which cannot be quantified called 'will'. It has been claimed that a computer, that mechanical mock-up of the synthesized molecular man we are anticipating, can make 'value judgements'. At best, that can only mean that in addition to its dictionary of facts it has been given a built-in book of rules which gives it a range of choice. That is not much different from saying that a thermostatic oven will make a 'good' cake if it is properly set and timed.

'Mind', 'will', 'judgement', 'inspiration', 'morality', 'kindness', 'tolerance', 'evil', 'brutality', all these are as much part of the human organism as 'respiration', 'nutriment', 'secretion', and 'excretion', but how do you express them in terms of the configuration of molecules or even of the measurable electronic excitation of the brain cells? The answer is, of course, you cannot, or, if you did, you would have to spell it out, not in terms of one prototype human but of each of thousands of millions. The physics and chemistry which can explain with increasing precision how the multicellular organism, Man, is constructed and how his sensory organs maintain his perception of the outside world cannot explain the combination of instinctive behaviour (like the reflexes of insects, birds and animals), which, with stored experience, is memory, but which also extends into imagination, inspiration, and decision-making, distinguishing, in greater or less degree, every individual personality.

In the same lecture as that in which he traced the construction of Man from the original cosmic hydrogen, Dr Beadle said:

> Especially after birth the information that is fed into the nervous system in massive amounts plays a large and important role in determining what we are. It includes a large input of cultural inheritance.[6]

6. op. cit.

Cells form tissues; tissues form organs; and organs form organisms in a state of cellular interdependence. Human beings form groups and groups form cultures, and cultures are an interdependence of shared experience and shared values. His own kind are part of Man's environment.

Culture has given us science. Our knowledge of science is not innate; it is acquired; it is accumulated and handed on; it is man-to-man information. By observation and cogitation; by experiments and measurements that have verified observation and confirmed speculation; by scepticism, with which each generation questions the tenets of the older masters; and by debate which challenges inconsistencies and relates one set of facts to another, science is establishing frames of reference, paradigms, which serve a contemporary purpose. There are some with an ambition (surely a non-molecular word?) to establish a framework which can encompass everything – the universe of the Cosmos, the behaviour of elemental particles, and the ions of the human brain which thinks about all these things. There are those who, less ambitiously, would settle for uniformity instead of unity – for patterns of consistency rather than a universal law.

Science is an adventure of the inquiring mind. It likes to speculate on what lies beyond the far horizons, but it also wants to beat the trail across those horizons – beyond the gravitational fences into space, for instance. Its markers for the trail are experimentally established facts, measurable and checkable. As each ridge is crossed, the theory is confirmed or replaced by fresh observations. And presently along the trail, come the covered wagons of advancing Mankind, acquiring the material advantages of science.

There is no finite vision because the inward eye of scientific imagination is not limited even by the speed of light.

Index